More information about this series at http://www.springer.com/series/7172

Editorial Board

Lisa Beck

Elliptic Regularity Theory

A First Course

Springer

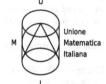
Unione
Matematica
Italiana

Lisa Beck
Institut für Mathematik
Universität Augsburg
Augsburg, Germany

ISSN 1862-9113   ISSN 1862-9121   (electronic)
Lecture Notes of the Unione Matematica Italiana
ISBN 978-3-319-27484-3   ISBN 978-3-319-27485-0   (eBook)
DOI 10.1007/978-3-319-27485-0

Library of Congress Control Number: 2015958734

Mathematics Subject Classification (2010): 35J47, 35B65, 49N60

Springer International Publishing AG Switzerland is part of Springer Science+Business Media
(www.springer.com)

Preface

We present a systematic and self-contained exposition of some aspects of regularity theory for elliptic equations of second order in divergence form. More specifically, our principal aim is to describe some fundamental techniques, which have been developed over the last decades, in order to tackle the problem of the optimal regularity of weak solutions. In doing so we will explain the dichotomy between the situation of one single equation with a weak solution with values in $\mathbb{R}$ (the "scalar case"), for which full regularity results may be established, and the situation of a system of several coupled equations with weak solutions with values in $\mathbb{R}^N$ (the "vectorial case"), in which discontinuities may actually arise.

These lecture notes are intended for graduate students who have a solid foundation in functional analysis and some familiarity of partial differential equations, although not necessarily with a specific background in the study of the regularity of weak solutions. For this reason, the main objective is not to present the most general results, as the extra technical details needed to obtain them often obscure the key underlying ideas. Instead, we prefer to introduce the different concepts and approaches to regularity in a way that is more suitable to nonexperts. In the scalar case, we cover the techniques of De Giorgi [15] and Moser [67] dating back to the late 1950s, which allow us to establish everywhere continuity of weak solutions under relatively mild assumptions on the equation. These techniques lay the foundation of what is nowadays referred to as *regularity theory*. However, in general, they cannot be extended to the vectorial case. This is not a technical issue but, rather, a structural one, for the vectorial situation is indeed fundamentally different to the scalar one. For instance, even if the system depends analytically on all variables, in the vectorial case there may exist weak solutions with discontinuities. However, all is not lost and one may still hope for partial regularity of weak solutions, that is, regularity outside of a set of measure zero, which, in general, is nonempty. The focus of these lecture notes is precisely this topic of partial regularity in the vectorial case. We give a survey of some more recent techniques, which have been developed since 1968 and which lead to proofs of partial regularity of weak solutions under reasonable assumptions. Specifically, we describe in detail the blow-up technique employed by Giusti and Miranda [41], the direct approach implemented by Giaquinta and Giusti [31] and by Ivert [48], and the method of $\mathcal{A}$-harmonic approximation used by Duzaar and Grotowski [21]. Furthermore, we address the possibility of finding an upper bound on the size of the singular set ("dimension reduction"), the set on which discontinuities of the weak solution may occur. Initially, the partial regularity proofs yield only that the singular set is of Lebesgue measure zero. Hence, it is a nontrivial issue to derive bounds on the Hausdorff dimension of the singular set strictly below the space dimension n. In this regard, we discuss some recent developments of Mingione [62, 63], which reflect the current state of the art in this field of research. Therefore, these lecture notes might also be of interest for researchers working on related topics.

In order to avoid a number of technicalities, we usually concentrate on (optimal) regularity results for equations or systems of equations for which weak solutions naturally belong to the Sobolev space $W^{1,2}$. However, the theory we present is general, in the sense that most of the results readily extend to the corresponding setting with weak solutions in the Sobolev space $W^{1,p}$ for an arbitrary $p \in (1, \infty)$. To the author's knowledge, such a treatise of the most recent regularity theory is not available in the classical monographs on the theory of partial differential equations, such as [30, 35, 39, 53, 57], and therefore, the lecture notes should serve as a complement for these textbooks. Finally, let us note that we restrict ourselves mainly to the theory for elliptic systems, even though the theory for convex variational problems is to a large extent very similar. Only for the sake of illustration of this similarity, we discuss briefly the essential steps of proof for two related regularity results for minimizers of (quadratic-growth) variational integrals, following the line of arguments for elliptic systems. We recommend, for instance, Giusti's monograph [40] for a in-depth presentation of some general techniques and results in this variational context, while for a review on the more recent progress, we refer to Mingione's survey papers [64, 65].

We now give a more detailed account of the content of these lecture notes, providing also a short description of the regularity results presented and some comments on their historical context. A deeper discussion of the features of the theory is postponed to the respective chapters.

Chapter 1 contains the prerequisites for the topic of elliptic regularity theory. We state the main properties of the relevant function spaces, namely, the Hölder, Morrey, Campanato, and (classical as well as fractional) Sobolev spaces. In particular, this chapter includes the continuous and compact embedding theorems of Sobolev, Morrey, and Rellich–Kondrachov, Poincaré-type inequalities, and several tools that are tailored to the application in the context of partial differential equations. Moreover, we here provide the measure theoretic arguments that allow to bound the Hausdorff dimension of the set of non-Lebesgue points of a Sobolev function (which is later on used in the dimension reduction for the set of singular points of a weak solution). These preliminaries are complemented by Appendix A, which contains some basic facts from functional analysis without proofs, and by Appendix B, which provides several technical iteration lemmata which are used throughout the regularity proofs.

Chapter 2 provides a short introduction to the concept of weak solutions for quasilinear elliptic equations (or systems of equations) in divergence form and motivates some elementary assumptions used throughout the lecture notes. Moreover, we comment on the connection to minimization problems of variational functionals via the Euler–Lagrange formalism.

Chapter 3 treats elliptic quasilinear equations under relatively general assumptions, that is, the scalar case. In the first part, De Giorgi's level-set technique from [15] is discussed. This technique allows us to prove that weak solutions to linear equations in divergence form with bounded, measurable coefficients are actually continuous. This solved the 19th of Hilbert's celebrated open problems presented at the International Congress of Mathematicians in 1900 in Paris and is considered an important milestone in elliptic regularity theory. We explain De Giorgi's technique in the general setting of Q-minimizers of variational functionals, which is a unified approach for obtaining the optimal regularity for minimizers of convex variational functionals and for weak solutions of elliptic equations simultaneously. By a careful analysis of the (super- and sub)level sets of Q-minimizers and associated Caccioppoli-type inequalities, first boundedness and then everywhere continuity of Q-minimizers is proved. The second part of the chapter outlines an alternative strategy of proof of this everywhere regularity result, which was developed by Moser [67] shortly after the original proof was published. It relies on a delicate iteration technique, using suitable test functions, Sobolev's embedding theorem, and a version of the John–Nirenberg lemma. This leads in a first step to boundedness and a Harnack inequality, whereby the supremum of a positive weak solution is bounded by its infimum. In a second step, the optimal regularity result is obtained, with the degree of regularity linked directly to the regularity of the equation.

Chapter 4 begins the discussion of the vectorial case, with an emphasis on the special case of elliptic systems that are linear in the gradient variable and, therefore, simpler to study. That the situation changes dramatically compared to the scalar one discussed in Chap. 3 was already observed by De Giorgi [16]. He constructed an example of such a system that admits a discontinuous weak solution, bringing to an end the efforts to find an extension of the everywhere regularity result to the vectorial case. We review his construction and a modification due to Giusti and Miranda [42], for which even an unbounded weak solution exists. However, in specific situations, one can still show that every weak solution is in a space of higher regularity (or even everywhere regular as in the scalar case), and we here give a brief summary of Morrey- and Campanato-type decay estimates and of the Schauder theory for linear systems. These are also the starting point for the investigation of partial regularity of weak solutions, which aims to show regularity of weak solutions outside of a set of singular points, together with a bound on the size of this exceptional set. In this regard, we illustrate, for the special type of system considered in this chapter, three different techniques, namely, the blow-up technique of Giusti and Miranda [41], the direct approach of Giaquinta and Giusti [31] and Ivert [48], and the method of $\mathcal{A}$-harmonic approximation of Duzaar and Grotowski [21]. The (common) relevant quantity is the excess, that is, the averaged mean-square deviation from the mean of the weak solution, for which decay estimates need to be established. This is achieved (at regular points) via comparison with a linearized system. It is precisely in the detail of how the linearization is implemented that the three approaches differ. However, they all lead to the same partial regularity result, namely, that weak solutions are continuous outside of a negligible set. Moreover, this regularity result comes along with a characterization of the set of points in which the weak solution is discontinuous, and this allows to deduce that its Hausdorff dimension does not exceed $n - 2$. Finally, we give a corresponding partial regularity result also for minimizers for quadratic variational integrals, via a modified version of the direct approach.

Chapter 5 continues to investigate the vectorial case, now for general quasilinear elliptic systems. The chapter begins with a proof that weak solutions are continuously differentiable outside of a negligible set and gives the characterization of this exceptional set. We follow the proof of Duzaar and Grotowski [21] via the method of $\mathcal{A}$-harmonic approximation, which appears to be more flexible than the other techniques. Moreover, we sketch the related partial regularity result for minimizers to convex variational integrals. In contrast to Chap. 4, where the regularity of the weak solution (or minimizer) and not of its gradient is considered, a nontrivial bound on the Hausdorff dimension of the singular set does not follow directly from its characterization, but requires further work. The first step toward the dimension reduction of the singular set was achieved by Mingione [62, 63]. The crucial idea is to show that weak solutions actually belong to a higher (fractional) Sobolev space, by taking advantage of the regularity of the system and in some instances possibly also of the weak solution itself. Furthermore, we include some refinements of the fractional differentiability estimates, which are based on a higher integrability result involving Gehring's lemma. Once the fractional differentiability is established, a bound on the Hausdorff dimension of the singular set follows in turn from measure theoretic arguments (similar to those in Chap. 4). In conclusion, we obtain also for quasilinear systems that, as basic intuition suggests, the more regular the system is, the more regular the weak solution is and the more the bound on the Hausdorff dimension of the singular set can be reduced. However, this result contains a slight drawback, namely, that we need to suppose that either the weak solution is a priori (Hölder) continuous or that the system does not depend explicitly on the weak solution, but only on its gradient. Therefore, we devote the last section of this chapter to a special partial regularity result for weak solutions, which was first given by Campanato [11] and which requires the assumption of low dimensions $n \leq 4$. This restriction enables us to apply a version of the direct approach, which deals with Morrey estimates, is based on a nonlinear comparison principle, and ends up with partial Hölder continuity of the weak solution, outside of a set of Hausdorff dimension not greater than $n - 2$.

I wish to conclude the preface with an acknowledgment. These lecture notes grew out of a course given at the University of Bonn in winter 2011/2012, where I was granted the opportunity to explain parts of my field of research to master and PhD students, and they were completed during a course for master students given at the University of Augsburg in summer 2015. I would like to thank the participants for their interest and comments, but I also express my sincere gratitude to my colleagues behind the scenes for encouragement and support. In particular, I would like to thank Julian Braun, Judith Campos Cordero, and Joseph Grotowski for valuable suggestions for improvements of parts of the manuscript.

Augsburg, Germany   Lisa Beck
October 2015

Contents

Chapter 1
Preliminaries

In this chapter we first recall several function spaces that will be relevant in order to tackle the questions of existence and regularity for (weak) solutions to elliptic problems. We further fix the notation and give some important properties, inequalities, and embedding theorems. Even though this theory would deserve to be developed in more detail and wider generality, we have decided to present the material in this chapter rather as a collection of the mathematical background. In particular, the proofs will be given only for some selected statements, which are of central interest in the course of these lecture notes or which are instructional and might serve as an illustration in order to understand the underlying concepts. For the other results and further information (also on functional analytic aspects) we refer to the literature, such as [1, 3, 24, 84, 85] and the references therein.

1.1 Function Spaces

In this section we give the definitions of several function classes, including the spaces of continuous and Hölder continuous functions, Lebesgue spaces, Morrey and Campanato spaces, and finally the classical as well as the fractional Sobolev spaces. Since the Morrey, Campanato and Sobolev spaces are defined in terms of weak differentiability, integrability and decay properties, they are perfectly adapted for the derivation of regularity criteria for functions obeying certain integral identities (such as weak solutions to systems of partial differential equations) or integral inequalities (such as minimizers of variational integrals), due to the embedding theorems for these spaces into more regular ones, which are given later on.

© Springer International Publishing Switzerland 2016
L. Beck, *Elliptic Regularity Theory*, Lecture Notes of the Unione
Matematica Italiana 19, DOI 10.1007/978-3-319-27485-0_1

1.1.1   Spaces of Continuous and Hölder Continuous Functions

We start by recalling the spaces of continuous functions and the spaces of continuous functions with compact support.

Definition 1.1 For a set Ω in $\mathbb{R}^n$ we define

(i) $C^0(\Omega, \mathbb{R}^N) = C(\Omega, \mathbb{R}^N)$ as the set of all continuous functions $f\colon \Omega \to \mathbb{R}^N$;

(ii) $C^0(\overline{\Omega}, \mathbb{R}^N) = C(\overline{\Omega}, \mathbb{R}^N)$ as the set of all functions in $C(\Omega, \mathbb{R}^N)$ which can be continuously extended to the closure $\overline{\Omega}$ of Ω;

(iii) $C_0(\Omega, \mathbb{R}^N)$ as the set of all functions $f \in C(\Omega, \mathbb{R}^N)$ with support $\operatorname{spt} f := \overline{\{x \in \Omega\colon f(x) \neq 0\}}$ compactly contained in Ω.

A function $f \in C(\Omega, \mathbb{R}^N)$ is not necessarily bounded on Ω. However, if f is bounded and uniformly continuous, then it can be uniquely extended up to the boundary, hence, it can actually be considered as a function in $C(\overline{\Omega}, \mathbb{R}^N)$.

Accordingly, one can define the spaces C^k as the set of all functions with continuous derivatives up to order $k \in \mathbb{N}_0$ (including also the function itself as derivative of order 0). For (partial) derivatives of higher order of a function f we shall use the notation $D^\beta f := D_1^{\beta_1} \ldots D_n^{\beta_n} f$, where $\beta = (\beta_1, \ldots, \beta_n) \in \mathbb{N}_0^n$ denotes a multiindex of length $|\beta| := \beta_1 + \ldots + \beta_n$. Furthermore, we shall write $D^k f$ in order to denote the vector $\{D^\beta f\}_{|\beta|=k}$ of the collection of all derivatives of order k.

Definition 1.2 Let Ω be an open set in $\mathbb{R}^n$ and $k \in \mathbb{N}$.

(i) $C^k(\Omega, \mathbb{R}^N)$ is the set of all functions $f\colon \Omega \to \mathbb{R}^N$ such that the partial derivatives $D^\beta f$ for all multiindices $\beta \in \mathbb{N}_0^n$ with $0 \leq |\beta| \leq k$ are continuous in Ω;

(ii) $C^k(\overline{\Omega}, \mathbb{R}^N)$ is the set of all functions in $C^k(\Omega, \mathbb{R}^N)$ whose derivatives up to order k are uniformly bounded in Ω and can be continuously extended to $\overline{\Omega}$;

(iii) $C^\infty(\Omega, \mathbb{R}^N) := \bigcap_{k \in \mathbb{N}} C^k(\Omega, \mathbb{R}^N)$ is the set of all functions $f\colon \Omega \to \mathbb{R}^N$ which are infinitely differentiable (smooth);

(iv) $C_0^k(\Omega, \mathbb{R}^N) := C^k(\Omega, \mathbb{R}^N) \cap C_0(\Omega, \mathbb{R}^N)$ and $C_0^\infty(\Omega, \mathbb{R}^N) := C^\infty(\Omega, \mathbb{R}^N) \cap C_0(\Omega, \mathbb{R}^N)$ are the sets of all functions in $C^k(\Omega, \mathbb{R}^N)$ and $C^\infty(\Omega, \mathbb{R}^N)$, respectively, with compact support in Ω.

We next recall the Hölder spaces, which are subspaces of the space of continuous functions C^k, for which slightly better regularity properties hold. These Hölder spaces roughly consist either of the well-known Hölder continuous functions in the case $k = 0$ or, in the case $k > 0$, of those functions whose k-th order partial derivatives are all Hölder continuous, respectively. In order to give the precise definition, we start by defining the Hölder semi-norm. Consider a number α in $(0, 1]$, a subset S of $\mathbb{R}^n$, and a function $f\colon S \to \mathbb{R}^N$.

The α-Hölder semi-norm of f in S is given by

$$[f]_{C^{0,\alpha}(S,\mathbb{R}^N)} := \sup_{x \neq y \in S} \left\{ \frac{|f(x) - f(y)|}{|x - y|^\alpha} \right\}.$$

Definition 1.3 Let Ω be an open set in $\mathbb{R}^n$, $k \in \mathbb{N}$, and $\alpha \in (0,1]$.

(i) $C^{0,\alpha}(\Omega, \mathbb{R}^N)$ is the set of all functions $f \in C(\Omega, \mathbb{R}^N)$ such that, for every compact set $K \subset \Omega$, $[f]_{C^{0,\alpha}(K,\mathbb{R}^N)}$ is finite;

(ii) $C^{0,\alpha}(\overline{\Omega}, \mathbb{R}^N)$ is the set of all bounded functions $f \in C(\overline{\Omega}, \mathbb{R}^N)$ such that $[f]_{C^{0,\alpha}(\overline{\Omega},\mathbb{R}^N)}$ is finite;

(iii) $C^{k,\alpha}(\Omega, \mathbb{R}^N)$ is the set of all functions $f \in C^k(\Omega, \mathbb{R}^N)$ such that $[D^\beta f]_{C^{0,\alpha}(K,\mathbb{R}^N)}$ is finite for every compact set $K \subset \Omega$ and every multiindex $\beta \in \mathbb{N}_0^n$ of length $|\beta| = k$;

(iv) $C^{k,\alpha}(\overline{\Omega}, \mathbb{R}^N)$ is the set of all functions $f \in C^k(\overline{\Omega}, \mathbb{R}^N)$ such that $[D^\beta f]_{C^{0,\alpha}(\overline{\Omega},\mathbb{R}^N)}$ is finite for every multiindex $\beta \in \mathbb{N}_0^n$ of length $|\beta| = k$.

Both the spaces of continuous and of Hölder continuous functions have good functional analytic properties, in the sense that they are complete normed spaces.

Theorem 1.4 *Let Ω be an open set in $\mathbb{R}^n$, $k \in \mathbb{N}_0$, and $\alpha \in (0,1]$. The spaces $C^k(\overline{\Omega}, \mathbb{R}^N)$ and $C^{k,\alpha}(\overline{\Omega}, \mathbb{R}^N)$ are Banach spaces, equipped with the norms*

$$\|f\|_{C^k(\overline{\Omega},\mathbb{R}^N)} := \sum_{0 \leq |\beta| \leq k} \sup_{x \in \overline{\Omega}} |D^\beta f(x)|,$$

$$\|f\|_{C^{k,\alpha}(\overline{\Omega},\mathbb{R}^N)} := \|f\|_{C^k(\overline{\Omega},\mathbb{R}^N)} + \sum_{|\beta|=k} [D^\beta f]_{C^{0,\alpha}(\overline{\Omega},\mathbb{R}^N)}.$$

Remarks 1.5

(i) To avoid confusion let us note that there are different conventions for the definition of the Hölder spaces, and sometimes the spaces $C^{0,\alpha}(\Omega, \mathbb{R}^N)$ are introduced as the spaces of bounded functions which are *uniformly* α-Hölder continuous in Ω.

(ii) For $\alpha = 1$, the space $C^{0,1}(\overline{\Omega}, \mathbb{R}^N)$ is the set of all Lipschitz continuous functions, that is, of all bounded functions $f : \overline{\Omega} \to \mathbb{R}^N$ which satisfy the Lipschitz condition

$$|f(x) - f(y)| \leq L|x - y| \qquad \text{for all } x, y \in \overline{\Omega}$$

with some constant $L \geq 0$. The best such (Lipschitz) constant L is given by the semi-norm $[f]_{C^{0,1}(\overline{\Omega},\mathbb{R}^N)}$.

(iii) The Hölder spaces are not separable (for example, the family of functions $f_{x_0}(x) := |x - x_0|^\alpha$ for $x_0 \in [0,1]$ is uncountable and every pair of functions f_{x_0}, f_{x_1} with $x_0 \neq x_1$ has distance greater than 1 in $C^{0,\alpha}([0,1])$). Note also that smooth functions are not dense in the Hölder spaces, and for $\alpha \in (0,1)$ the closure of $C^\infty(\Omega, \mathbb{R}^N)$ in $C^{0,\alpha}(\Omega, \mathbb{R}^N)$ is known as the *little Hölder space* $c^{0,\alpha}(\Omega, \mathbb{R}^N)$, while for $\alpha = 1$ this closure is the space $C^1(\Omega, \mathbb{R}^N)$.

(iv) If $0 < \alpha_1 \leq \alpha_2 \leq 1$ and $k \in \mathbb{N}_0$, then we obviously have the continuous embeddings

$$C^{k,1}(\overline{\Omega}, \mathbb{R}^N) \subset C^{k,\alpha_2}(\overline{\Omega}, \mathbb{R}^N) \subset C^{k,\alpha_1}(\overline{\Omega}, \mathbb{R}^N) \subset C^k(\overline{\Omega}, \mathbb{R}^N)$$

(whereas the validity of the inclusions $C^{k+1}(\overline{\Omega}, \mathbb{R}^N) \subset C^{k,1}(\overline{\Omega}, \mathbb{R}^N)$ depends on the choice of Ω, see [39, p. 53] for an example of a domain, for which this embedding fails).

In relation with the last remark, we also recall the Arzelà–Ascoli compactness theorem, which provides a sufficient condition that, for a given sequence of continuous functions on a compact set, there exists of a uniformly convergent subsequence.

Theorem 1.6 (Arzelà–Ascoli) *Let Ω be a bounded set in $\mathbb{R}^n$. A sequence $(f_j)_{j \in \mathbb{N}}$ of functions in $C(\overline{\Omega})$ has a uniformly convergent subsequence if they are uniformly bounded and equicontinuous, that is if $\sup_{j \in \mathbb{N}} \sup_{x \in \Omega} |f_j(x)| < \infty$ and if for every $\varepsilon > 0$ there exists a $\delta > 0$ such that for all $j \in \mathbb{N}$ there holds*

$$|f_j(x) - f_j(y)| < \varepsilon \qquad \text{for all } x, y \in \overline{\Omega} \text{ with } |x - y| < \delta \,.$$

1.1.2 Lebesgue Spaces

We next recall the Lebesgue spaces $L^p(\Omega)$, which consist of all measurable functions that are integrable to some power $p \in [1, \infty]$ over a given measurable set Ω.

Definition 1.7 Let Ω be a measurable set in $\mathbb{R}^n$ and let $p \in [1, \infty]$. We denote by $L^p(\Omega, \mathbb{R}^N)$ the *Lebesgue space* of (equivalence classes of) measurable functions $f \colon \Omega \to \mathbb{R}^N$ such that

$$\|f\|_{L^p(\Omega, \mathbb{R}^N)} := \begin{cases} \left(\int_\Omega |f|^p \, dx \right)^{\frac{1}{p}} & \text{if } 1 \leq p < \infty \\ \operatorname{ess\,sup}_\Omega |f| & \text{if } p = \infty \end{cases} \tag{1.1}$$

is finite. We further denote by $L^p_{\mathrm{loc}}(\Omega, \mathbb{R}^N)$ the set of all functions belonging to $L^p(O, \mathbb{R}^N)$ for every open set $O \Subset \Omega$. For scalar-valued functions (i.e. for the case $N = 1$), we write $L^p(\Omega)$ instead of $L^p(\Omega, \mathbb{R})$ and $L^p_{\mathrm{loc}}(\Omega)$ instead of $L^p_{\mathrm{loc}}(\Omega, \mathbb{R})$.

Let us explain briefly how and why equivalence classes of measurable functions enter into the previous definition. One could define a space $\mathcal{L}^p(\Omega, \mathbb{R}^N)$ as the set of all measurable function such that (1.1) is finite. This is a vector-space, but it does not satisfy the Hausdorff separation axioms since functions that differ only on a Lebesgue null set are not distinguishable by (1.1). This obstruction is easily resolved by first taking the subspace of all functions in $\mathcal{L}^p(\Omega, \mathbb{R}^N)$ for which (1.1) is equal to 0, and by then building the space $L^p(\Omega, \mathbb{R}^N)$ as the quotient space of $\mathcal{L}^p(\Omega, \mathbb{R}^N)$ by this subspace. Consequently, the elements of $L^p(\Omega, \mathbb{R}^N)$ are equivalence classes of measurable functions that differ only on a set of Lebesgue measure zero. In a sloppy way, we will usually speak of functions (and mean the corresponding equivalence classes). The choice of good representatives will be of importance later when properties not only outside of a negligible set but on the full domain are required (such as continuity).

Remark 1.8 If f belongs to $L^p(\Omega, \mathbb{R}^N)$, for some $p \in [1, \infty)$, then the super-level sets of $|f|$ at height $\ell > 0$ are estimated in measure by

$$\mathcal{L}^n\left(\{x \in \Omega \colon |f(x)| > \ell\}\right) \leq \ell^{-p}\|f\|_{L^p(\Omega, \mathbb{R}^N)}^p. \qquad (1.2)$$

This property can be used to define the *weak Lebesgue space* $L_w^p(\Omega, \mathbb{R}^N)$ as the space of all (equivalence classes of) measurable functions $f \colon \Omega \to \mathbb{R}^N$ such that

$$\|f\|_{L_w^p(\Omega, \mathbb{R}^N)}^p := \sup_{\ell > 0} \ell^p \mathcal{L}^n\left(\{x \in \Omega \colon |f(x)| > \ell\}\right) \quad \text{is finite.}$$

Before addressing some important features of the L^p-spaces, we remind the reader of some important convergence theorems from general measure theory, see e.g. [25, Chapter 1.3].

Theorem 1.9 (Fatou) *If $(f_j)_{j \in \mathbb{N}}$ is a sequence of non-negative, measurable functions on $\mathbb{R}^n$, then there holds*

$$\int_{\mathbb{R}^n} \liminf_{j \to \infty} f_j \, dx \leq \liminf_{j \to \infty} \int_{\mathbb{R}^n} f_j \, dx.$$

Theorem 1.10 (Monotone convergence) *If $(f_j)_{j \in \mathbb{N}}$ is a sequence of non-negative, measurable functions on $\mathbb{R}^n$, which are monotonically non-decreasing in j (that is, we have $f_j \leq f_{j+1}$ for every $j \in \mathbb{N}$), then there holds*

$$\int_{\mathbb{R}^n} \lim_{j \to \infty} f_j \, dx = \lim_{j \to \infty} \int_{\mathbb{R}^n} f_j \, dx.$$

Theorem 1.11 (Variant of the dominated convergence) *Let $(g_j)_{j \in \mathbb{N}}$ be a sequence of non-negative functions which converges strongly to a function g in $L^1(\mathbb{R}^n)$. If $(f_j)_{j \in \mathbb{N}}$ is a sequence of functions in $L^1(\mathbb{R}^n)$, which satisfies*

$$|f_j| \leq g_j \quad \text{for all } j \in \mathbb{N} \qquad \text{and} \qquad f_j(x) \to f(x) \quad \mathcal{L}^n\text{-almost everywhere}$$

for a function $f \in L^1(\mathbb{R}^n)$, then $f_j \to f$ strongly in $L^1(\mathbb{R}^n)$.

Returning to the discussion of L^p-spaces, we start by noting that the space $C_0^\infty(\Omega, \mathbb{R}^N)$ of smooth and compactly supported functions is dense in $L^p(\Omega, \mathbb{R}^N)$ for $p \in [1, \infty)$ (while for $p = \infty$ this is obviously false), as a consequence of Lusin's theorem and the absolute continuity of the integral. Concerning the pointwise behavior of functions, we know for example that every smooth function coincides at every point with the function that is obtained by taking the limit of infinitesimal averages about that point. This fact continues to be true also for functions in the Lebesgue spaces, but, due to the previous discussion that modifications on sets of measure zero do not change the equivalence class of the function and that there are equivalence classes which do not contain a continuous function as its representative, we have to accept the limitation that this coincidence holds only outside of a set of Lebesgue measure zero. Before stating this result, known as the classical Lebesgue differentiation theorem, let us introduce a notation for the mean values of a Lebesgue function. For a given measurable set $S \subset \mathbb{R}^n$ we denote by $\mathcal{L}^n(S) = |S|$ its n-dimensional Lebesgue measure. Furthermore, if $f \in L^1(S, \mathbb{R}^N)$ and $0 < |S| < \infty$, we denote the average of f in S by

$$(f)_S := \fint_S f \, dx := \frac{1}{|S|} \int_S f \, dx \,.$$

With this notation the aforementioned differentiation theorem reads as follows.

Theorem 1.12 (Lebesgue differentiation theorem) *Let $f \in L^1_{\mathrm{loc}}(\mathbb{R}^n)$. Then, for almost every $x_0 \in \mathbb{R}^n$, we have*

$$\lim_{\varrho \searrow 0} \fint_{B_\varrho(x_0)} |f(x) - f(x_0)| \, dx = 0$$

(such a point is called a Lebesgue point *of f), and in particular there holds*

$$\lim_{\varrho \searrow 0} \fint_{B_\varrho(x_0)} f(x) = f(x_0) \,.$$

As a further benefit, this limit of infinitesimal averages over balls can be employed to choose a canonical representative in each Lebesgue class. It further allows to deduce, as a direct corollary, a version for L^p-spaces, cp. [25, Chapter 1.7, Corollary 1].

Corollary 1.13 *Let $p \in [1, \infty)$ and $f \in L^p_{loc}(\mathbb{R}^n)$. Then, for almost every $x_0 \in \mathbb{R}^n$, we have*

$$\lim_{\varrho \searrow 0} \fint_{B_\varrho(x_0)} |f(x) - f(x_0)|^p \, dx = 0$$

(such a point is called a p-Lebesgue point of f).

We continue by recalling some fundamental inequalites.

Theorem 1.14 (Jensen's inequality) *Let $g \in C(\mathbb{R})$ be a convex function and $f \in L^1(\Omega)$ for some measurable set $\Omega \subset \mathbb{R}^n$ with $0 < |\Omega| < \infty$. Then we have*

$$g\left(\fint_\Omega f \, dx \right) \le \fint_\Omega g \circ f \, dx.$$

For $p \in [1, \infty]$ we introduce the conjugate exponent $p' \in [1, \infty]$ defined via the equation $1/p + 1/p' = 1$ (with the convention $1/\infty = 0$). For $p \in (1, \infty)$, we first recall *Young's inequality*

$$ab \le \frac{a^p}{p} + \frac{b^{p'}}{p'} \tag{1.3}$$

for all $a, b \in \mathbb{R}_0^+$ (which is a direct consequence of the convexity of the exponential function). In the course of these lectures notes, we will frequently apply Young's inequality, in an ε-version, to the product $ab = (a\varepsilon^{1/p})(b\varepsilon^{-1/p})$ for some positive number ε. This allows, on the one hand, to make the contribution of a^p arbitrarily small or, on the other hand, to have a balancing effect. The latter benefit is used for instance in the proof of the following important inequality (which is known as Cauchy–Schwarz inequality for $p = p' = 2$).

Proposition 1.15 (Hölder's inequality) *Let $p \in [1, \infty]$, $f \in L^p(\Omega)$, and $g \in L^{p'}(\Omega)$. Then the product fg belongs to the space $L^1(\Omega)$ with*

$$\int_\Omega |fg| \, dx \le \|f\|_{L^p(\Omega)} \|g\|_{L^{p'}(\Omega)}.$$

Proof We first note that this inequality is obvious whenever $\|f\|_{L^p(\Omega)} = 0$ or $\|g\|_{L^{p'}(\Omega)} = 0$. Since the inequality is also trivial for $p \in \{1, \infty\}$, we may further assume $p \in (1, \infty)$. For an arbitrary $\varepsilon > 0$, Young's inequality (1.3) implies

$$\int_\Omega |fg| \, dx \le \frac{\varepsilon}{p} \int_\Omega |f|^p \, dx + \frac{\varepsilon^{-p'/p}}{p'} \int_\Omega |g|^{p'} \, dx$$

which gives the claim for the specific choice $\varepsilon = \|f\|_{L^p(\Omega)}^{1-p}\|g\|_{L^{p'}(\Omega)}$, via the identity $1/p + 1/p' = 1$.   □

Remarks 1.16

(i) Inserting piecewise constant functions on an interval in $\mathbb{R}$ we recover the discrete version of Hölder's inequality for vectors $a, b \in \mathbb{R}^k$ and $p \in (1, \infty)$

$$\sum_{\ell=1}^k |a_\ell b_\ell| \le \Big(\sum_{\ell=1}^k a_\ell^p \Big)^{\frac{1}{p}} \Big(\sum_{\ell=1}^k b_\ell^{p'} \Big)^{\frac{1}{p'}}.$$

(ii) Hölder's inequality implies the following interpolation inequality: for $f \in L^{p_1}(\Omega) \cap L^{p_2}(\Omega)$ with $1 \le p_1 < p_2 \le \infty$, we have $f \in L^q(\Omega)$ for all $q \in (p_1, p_2)$ with

$$\|f\|_{L^q(\Omega)} \le \|f\|_{L^{p_1}(\Omega)}^{\theta} \|f\|_{L^{p_2}(\Omega)}^{1-\theta},$$

where $\theta \in (0, 1)$ is determined via the equation $1/q = \theta/p_1 + (1 - \theta)/p_2$ (with the convention $\theta = p_1/q$ for $p_2 = \infty$).

(iii) We get the inclusions $L^q(\Omega) \subset L^p(\Omega)$ for all $p, q \in [1, \infty]$ with $q \ge p$ provided that Ω is of finite $\mathcal{L}^n$-measure. In particular, for every ball $B_r(x_0) \subset \mathbb{R}^n$ and $f \in L^q(B_r(x_0))$, we have the inequality

$$\|f\|_{L^p(B_r(x_0))} \le c(n) r^{n(\frac{1}{p} - \frac{1}{q})} \|f\|_{L^q(B_r(x_0))}.$$

By an inductive argument we find the generalized Hölder inequality for more than two functions:

Corollary 1.17 *Let $m \in \mathbb{N}$ and let $f_j \in L^{p_j}(\Omega)$ with $p_j \in [1, \infty]$ for $j \in \{1, \ldots, m\}$. If $\sum_{1 \le j \le m} p_j^{-1} = 1$, then the product $\prod_{1 \le j \le m} f_j$ belongs to $L^1(\Omega)$ with*

$$\int_\Omega |f_1 \ldots f_m| \, dx \le \prod_{1 \le j \le m} \|f_j\|_{L^{p_j}(\Omega)}.$$

We conclude this section with some statements on the spaces $L^p(\Omega, \mathbb{R}^N)$ as function spaces, giving in particular a sketch of proof for their completeness and stating a simple criterion for compactness in the weak topology (derived from the duality relations).

Theorem 1.18 *Endowed with the norm $\|\cdot\|_{L^p(\Omega, \mathbb{R}^N)}$ defined in (1.1), the Lebesgue spaces $L^p(\Omega, \mathbb{R}^N)$ are Banach spaces for all $p \in [1, \infty]$.*

Proof

Step 1: (1.1) defines a norm. We obviously have point separation (i.e. $\|f\|_{L^p(\Omega,\mathbf{R}^N)} = 0$ implies $f = 0$) and absolute homogeneity (i.e. $\|\lambda f\|_{L^p(\Omega,\mathbf{R}^N)} = |\lambda|\|f\|_{L^p(\Omega,\mathbf{R}^N)}$ for all $\lambda \in \mathbf{R}$), hence only the triangle inequality

$$\|f + g\|_{L^p(\Omega,\mathbf{R}^N)} \leq \|f\|_{L^p(\Omega,\mathbf{R}^N)} + \|g\|_{L^p(\Omega,\mathbf{R}^N)}$$

for $f, g \in L^p(\Omega, \mathbf{R}^N)$ and $p \in [1, \infty]$ remains to be checked. This inequality is usually referred to as *Minkowski's inequality* and is proved as follows: for $p \in \{1, \infty\}$ it is a direct consequence of the pointwise triangle inequality for the absolute value. For $p \in (1, \infty)$ this standard triangle inequality implies in a first step that the function $f + g$ belongs to $L^p(\Omega, \mathbf{R}^N)$ (more precisely, it proves the desired inequality up to a multiplicative factor). This justifies the application of Hölder's inequality from Proposition 1.15 with exponents $p/(p-1)$ and p, which yields

$$\|f + g\|^p_{L^p(\Omega,\mathbf{R}^N)} \leq \int_\Omega |f + g|^{p-1}(|f| + |g|)\,dx$$
$$\leq \|f + g\|^{p-1}_{L^p(\Omega,\mathbf{R}^N)}\big(\|f\|_{L^p(\Omega,\mathbf{R}^N)} + \|g\|_{L^p(\Omega,\mathbf{R}^N)}\big),$$

and the desired inequality follows immediately.

Step 2: Completeness of $L^p(\Omega, \mathbf{R}^N)$ (known as Riesz–Fischer theorem*).* We need to show that every Cauchy sequence $(f_j)_{j\in\mathbb{N}}$ in $L^p(\Omega, \mathbf{R}^N)$ converges to a limit in $L^p(\Omega, \mathbf{R}^N)$. Since Cauchy sequences cannot have more than one cluster point, it is sufficient to prove that an arbitrary subsequence of $(f_j)_{j\in\mathbb{N}}$ converges in $L^p(\Omega, \mathbf{R}^N)$. After possibly passing to a subsequence, we may suppose that $\|f_j - f_\ell\|_{L^p(\Omega,\mathbf{R}^N)} \leq 2^{-i}$ holds for all $j, \ell \geq i$ and therefore, we may also work under the assumption $\sum_{j=1}^\infty \|f_j - f_{j+1}\|_{L^p(\Omega,\mathbf{R}^N)} < \infty$. We then define a sequence of auxiliary functions

$$h_\ell(x) := \sum_{j=1}^\ell |f_j(x) - f_{j+1}(x)|$$

for $\ell \in \mathbb{N}$, and we set $h(x) := \lim_{\ell\to\infty} h_\ell(x)$, when the limit is well-defined. It is easy to check that this is the case for all $x \in \Omega \setminus S$, where S is a set of Lebesgue measure zero (for $p \in [1, \infty)$ this is a consequence of Theorem 1.9 of Fatou and Minkowski's inequality, while for $p = \infty$ the function h is even uniformly bounded in those points). Thus, also the sequences $(f_j(x))_{j\in\mathbb{N}}$ are Cauchy sequences in $\mathbf{R}^N$ for all $x \in \Omega \setminus S$ and hence, they converge to a limit $f(x)$. Obviously, the set S does depend on the choice of representatives, but for our purposes it is sufficiently to know that the pointwise limits exist outside of some set of Lebesgue measure zero. Now we extend f by 0 in this exceptional set S, and we observe that the resulting function f is again measurable, as the limit of measurable functions. We finally need to verify that f is indeed the desired L^p-limit function of the sequence $(f_j)_{j\in\mathbb{N}}$. For $p = \infty$, this is true, since the inequality $|f(x) - f_\ell(x)| = \lim_{j\to\infty} |f_j(x) - f_\ell(x)| \le 2^{-\ell}$ outside of a negligible set already implies $\|f - f_\ell\|_{L^\infty(\Omega,\mathbb{R}^N)} \to 0$ as $\ell \to \infty$. For $p \in [1,\infty)$, the convergence $\|f - f_\ell\|_{L^p(\Omega,\mathbb{R}^N)} \to 0$ is again a consequence of Fatou's Lemma, since we have

$$\int_\Omega |f - f_\ell|^p \, dx \le \liminf_{j\to\infty} \int_\Omega |f_j - f_\ell|^p \, dx$$

$$= \liminf_{j\to\infty} \|f_j - f_\ell\|_{L^p(\Omega,\mathbb{R}^N)}^p \to 0 \qquad \text{as } \ell \to \infty. \qquad \square$$

Remark 1.19 Similarly as in Remark 1.16 (i) there is a discrete version of Minkowski's inequality for vectors $a, b \in \mathbb{R}^k$ and $p \in [1,\infty)$:

$$\left(\sum_{\ell=1}^k (a_\ell + b_\ell)^p\right)^{\frac{1}{p}} \le \left(\sum_{\ell=1}^k a_\ell^p\right)^{\frac{1}{p}} + \left(\sum_{\ell=1}^k b_\ell^p\right)^{\frac{1}{p}}.$$

Remark 1.20 The space $L^2(\Omega,\mathbb{R}^N)$ is a Hilbert space, where the inner product is given by

$$\langle f, g\rangle_{L^2(\Omega,\mathbb{R}^N)} := \int_\Omega f \cdot g \, dx$$

for all functions $f, g \in L^2(\Omega,\mathbb{R}^N)$.

From Hölder's inequality it is further obvious that, for every function $g \in L^{p'}(\Omega,\mathbb{R}^N)$, one can define a linear functional L_g on $L^p(\Omega,\mathbb{R}^N)$ via integration against g, that is

$$L_g(f) := \int_\Omega fg \, dx \qquad \text{for all } f \in L^p(\Omega,\mathbb{R}^N).$$

It is easy to verify that the operator L (which maps g to the functional L_g) is an isometric isomorphism of $L^{p'}(\Omega,\mathbb{R}^N)$ into a subspace of the dual space of $L^p(\Omega,\mathbb{R}^N)$, which is denoted by $(L^p(\Omega,\mathbb{R}^N))^*$. It turns out that the range of L is all of $(L^p(\Omega,\mathbb{R}^N))^*$ for $p \in [1,\infty)$, while for $p = \infty$ the dual space $(L^\infty(\Omega,\mathbb{R}^N))^*$ is in general strictly larger than $L^1(\Omega,\mathbb{R}^N)$. This relation between the spaces $L^p(\Omega,\mathbb{R}^N)$ and $L^{p'}(\Omega,\mathbb{R}^N)$ also justifies to call the conjugate exponent p' of p the *dual* exponent.

Theorem 1.21 *Let $p \in [1,\infty)$. Then the space $(L^p(\Omega,\mathbb{R}^N))^*$ is isomorphic to $L^{p'}(\Omega,\mathbb{R}^N)$. In particular, $L^p(\Omega,\mathbb{R}^N)$ is reflexive for $p \in (1,\infty)$.*

The reflexivity of the L^p-spaces, for $p \in (1, \infty)$, immediately implies a very useful compactness property with respect to the weak topology in L^p, due to Theorem A.10, which states the equivalence between reflexivity of a general Banach space and weak precompactness of bounded sets in this space.

Finally, we state Gehring's lemma, which is based on Gehring's paper [29] on higher integrability of quasiconformal mappings and was later generalized by Giaquinta and Modica, see [37, Proposition 5.1], to the following statement, which allows to establish higher integrability of an integrable function.

Theorem 1.22 (Gehring; Giaquinta and Modica) *Let* $f \in L^1(B_R(x_0))$, $\sigma \in (0, 1)$, *and* $m \in (0, 1)$. *Suppose that there exist a constant* A *and a function* $g \in L^q(B_R(x_0))$ *for some* $q > 1$ *such that for all balls* $B_\varrho(y) \Subset B_R(x_0)$ *there holds*

$$\fint_{B_{\sigma\varrho}(y)} |f| \, dx \le A \Big(\fint_{B_\varrho(y)} |f|^m \, dx \Big)^{\frac{1}{m}} + \fint_{B_\varrho(y)} |g| \, dx .$$

Then there exists an exponent $p \in (1, q]$ *depending only on* A, m *and* n *such that* $f \in L^p_{\text{loc}}(B_R(x_0))$. *Moreover, for every* $\tau \in (0, 1)$ *we have*

$$\Big(\fint_{B_{\tau R}(x_0)} |f|^p \, dx \Big)^{\frac{1}{p}} \le K(A, m, n, \tau) \Big[\fint_{B_R(x_0)} |f| \, dx + \Big(\fint_{B_R(x_0)} |g|^p \, dx \Big)^{\frac{1}{p}} \Big] .$$

For $g = 0$ this can be viewed as a self-improvement property of (uniform) reverse-Hölder inequalities. The heuristic reason, why the higher integrability and the corresponding estimate is true, lies essentially in the fact that, with the hypothesis being valid uniformly on all interior balls, the function under consideration cannot develop too big concentrations of the norm on small sets. However, the proof of Theorem 1.22 is quite technical and involved, and for details we refer to the proof of [40, Theorem 6.6] (relying essentially on Calderón–Zygmund decompositions, covering arguments and estimates for super-level sets).

1.1.3 Morrey and Campanato Spaces

We next introduce the Morrey and Campanato spaces which are subspaces of the L^p-spaces with a finer structure, in the sense that they allow for an upper bound on the scaling of the L^p-norm in small balls in terms of powers of the radii of these balls.

Definition 1.23 Let Ω be an open set in $\mathbb{R}^n$, $p \in [1, \infty)$ and $\lambda \geq 0$.

(i) We denote by $L^{p,\lambda}(\Omega, \mathbb{R}^N)$ the *Morrey space* of all functions $f \in L^p(\Omega, \mathbb{R}^N)$ such that

$$\|f\|^p_{L^{p,\lambda}(\Omega,\mathbb{R}^N)} := \sup_{x_0 \in \overline{\Omega}, \varrho > 0} \min\{\varrho, 1\}^{-\lambda} \int_{\Omega(x_0, \varrho)} |f|^p \, dx \qquad (1.4)$$

is finite. Here we have employed the notation $\Omega(x_0, \varrho) := B_\varrho(x_0) \cap \Omega$.

(ii) We denote by $\mathcal{L}^{p,\lambda}(\Omega, \mathbb{R}^N)$ the *Campanato space* of all functions $f \in L^p(\Omega, \mathbb{R}^N)$ such that

$$[f]^p_{\mathcal{L}^{p,\lambda}(\Omega,\mathbb{R}^N)} := \sup_{x_0 \in \overline{\Omega}, \varrho > 0} \varrho^{-\lambda} \int_{\Omega(x_0, \varrho)} |f - (f)_{\Omega(x_0, \varrho)}|^p \, dx$$

is finite.

As a direct consequence of the completeness of the Lebesgue spaces, stated in Theorem 1.18, we observe that for all $p \in [1, \infty)$ and $\lambda \geq 0$ the Morrey spaces $L^{p,\lambda}(\Omega, \mathbb{R}^N)$ are Banach spaces, endowed with the norm $\|\cdot\|_{L^{p,\lambda}(\Omega,\mathbb{R}^N)}$ defined in (1.4). Similarly, the Campanato spaces $\mathcal{L}^{p,\lambda}(\Omega, \mathbb{R}^N)$ are Banach spaces, endowed with the norm $\|\cdot\|_{\mathcal{L}^{p,\lambda}(\Omega,\mathbb{R}^N)} := [\cdot]_{\mathcal{L}^{p,\lambda}(\Omega,\mathbb{R}^N)} + \|\cdot\|_{L^p(\Omega,\mathbb{R}^N)}$.

Remarks 1.24

(i) For bounded domains one usually uses the (equivalent) definition of Morrey spaces, where the factor $\varrho^{-\lambda}$ instead of $\min\{\varrho, 1\}^{-\lambda}$ is used (whereas for unbounded domains such a condition would in general not be sufficient to guarantee global L^p-integrability).

(ii) From the definition it is clear that, in order to verify that an $L^p(\Omega, \mathbb{R}^N)$-function belongs to $L^{p,\lambda}(\Omega, \mathbb{R}^N)$ or to $\mathcal{L}^{p,\lambda}(\Omega, \mathbb{R}^N)$, one needs to check the two conditions only for small radii $\varrho < \varrho_0$ for some fixed, positive number ϱ_0 (such as diam Ω for a bounded set Ω).

(iii) In view of Remark 1.16 (iii) we have for a bounded set Ω the inclusions

$$L^{q,\mu}(\Omega, \mathbb{R}^N) \subset L^{p,\lambda}(\Omega, \mathbb{R}^N) \quad \text{and} \quad \mathcal{L}^{q,\mu}(\Omega, \mathbb{R}^N) \subset \mathcal{L}^{p,\lambda}(\Omega, \mathbb{R}^N),$$

whenever the inequalities $q \geq p$ and $(n - \lambda)/p \geq (n - \mu)/q$ are satisfied.

(iv) If f belongs to $L^{p,\lambda}(\Omega, \mathbb{R}^N)$, then the super-level sets of $|f|$ at height $\ell > 0$ in any $\Omega(x_0, \varrho)$ are estimated by

$$\mathcal{L}^n(\{x \in \Omega(x_0, \varrho): |f(x)| > \ell\}) \leq \ell^{-p} \min\{\varrho, 1\}^\lambda \|f\|^p_{L^{p,\lambda}(\Omega,\mathbb{R}^N)}.$$

Similarly as for the weak Lebesgue spaces introduced in Remark 1.8, one can define the *weak Morrey space* $L_w^{p,\lambda}(\Omega, \mathbb{R}^N)$ as the set of all functions $f \in L_w^p(\Omega, \mathbb{R}^N)$ such that

$$\sup_{\ell>0, x_0 \in \overline{\Omega}, \varrho > 0} \ell^p \min\{\varrho, 1\}^{-\lambda} \mathcal{L}^n(\{x \in \Omega(x_0, \varrho) : |f(x)| > \ell\}) \quad \text{is finite.}$$

Moreover, we want to comment on several equivalence relations for Morrey and Campanato spaces, for the particular case that Ω is bounded and sufficiently regular, in the sense that Ahlfor's regularity condition on Ω holds true. This condition requires

$$|\Omega(x_0, \varrho)| \geq A\varrho^n \qquad \text{for all } x_0 \in \overline{\Omega} \text{ and every } \varrho \leq \operatorname{diam}(\Omega), \qquad (1.5)$$

for some constant $A > 0$. Geometrically, this means that the domain has no exterior cusps and that the measure of $\Omega(x_0, \varrho)$ can be estimated from below in terms of the factor ϱ^n. This condition hence guarantees that the factor $\varrho^{-\lambda}$ in the definition of Morrey and Campanato spaces can be replaced by $|\Omega(x_0, \varrho)|^{-\lambda/n}$. We further emphasize that this condition is for example satisfied if Ω has a Lipschitz boundary, which can be thought of as locally being the graph of a Lipschitz continuous function. More precisely, we say that a bounded, open subset Ω in $\mathbb{R}^n$ has a Lipschitz boundary if for every boundary point $y \in \partial\Omega$ there exist a radius $r(y) > 0$ and a bijection $b_y \colon B_{r(y)}(y) \to B_1$ such that both functions b_y and b_y^{-1} are Lipschitz continuous and such that the identities $b_y(\partial\Omega \cap B_{r(y)}(y)) = B_1 \cap (\mathbb{R}^{n-1} \times \{0\})$ and $b_y(\Omega \cap B_{r(y)}(y)) = B_1 \cap (\mathbb{R}^{n-1} \times \mathbb{R}^+)$ hold (in particular, Ω and $\mathbb{R}^n \setminus \Omega$ are situated locally exactly at one side of the boundary $\partial\Omega$).

Remarks 1.25 Let Ω be a bounded, open set in $\mathbb{R}^n$ for which the Ahlfor's regularity condition (1.5) is fulfilled for some $A > 0$, and let $p \in [1, \infty)$. Then we have the following equivalences:

(i) For $\lambda \in [0, n)$ we have $\mathcal{L}^{p,\lambda}(\Omega, \mathbb{R}^N) = L^{p,\lambda}(\Omega, \mathbb{R}^N)$, and for $\lambda = 0$ they coincide with the standard Lebesgue space $L^p(\Omega, \mathbb{R}^N)$, see [35, Proposition 5.4].

(ii) For $\lambda = n$ the Morrey spaces $L^{p,n}(\Omega, \mathbb{R}^N)$ are all equivalent and coincide with the space $L^\infty(\Omega, \mathbb{R}^N)$, while the Campanato spaces $\mathcal{L}^{p,n}(\Omega, \mathbb{R}^N)$ coincide with the space $BMO(\Omega, \mathbb{R}^N)$, that is, with the space of functions of bounded mean oscillation, which was introduced by John and Nirenberg in [50] in 1961. This space is of special interest and characterized below; in particular, it is smaller than any Lebesgue space $L^p(\Omega, \mathbb{R}^N)$ with $p < \infty$, but it still contains $L^\infty(\Omega, \mathbb{R}^N)$ as a strict subspace.

(iii) For $\lambda > n$ we have for the Morrey spaces $L^{p,\lambda}(\Omega, \mathbb{R}^N) \simeq \{0\}$ in view of Lebesgue's differentiation Theorem 1.12. Concerning the Campanato spaces we need to distinguish two cases: for $\lambda \in (n, n+p]$ the spaces

$\mathcal{L}^{p,\lambda}(\Omega, \mathbb{R}^N)$ describe an integral characterization of Hölder continuous functions (where the Hölder exponent is given by $(\lambda - n)/p$), which was developed by Campanato [8] and simultaneously by Meyers [60]. Because of its fundamental relevance for regularity theory, this result is discussed in detail below in Theorem 1.27. For $\lambda > n + p$ and Ω connected, we finally have $\mathcal{L}^{p,\lambda}(\Omega, \mathbb{R}^N) \simeq \{\text{constants}\}$ (which can be seen as an analogue of the fact that all Hölder continuous functions with Hölder exponent greater than 1 are actually constant).

We next state, in the special situation of cubes, a characterization of the Campanato spaces $\mathcal{L}^{p,n}$. For its quite technical and involved proof, which relies heavily on cube decompositions and Calderón–Zygmund-type arguments, we refer to [35, Chapter 6.3.1].

Theorem 1.26 (John–Nirenberg) *Let Q_0 be a cube in $\mathbb{R}^n$ and consider a measurable function $f\colon Q_0 \to \mathbb{R}^N$. Then the following statements are equivalent:*

(i)   *There holds $f \in \mathcal{L}^{p,n}(Q_0, \mathbb{R}^N)$ for some $p \in [1, \infty)$.*
(ii) *There holds $f \in \mathcal{L}^{1,n}(Q_0, \mathbb{R}^N)$.*
(iii) *There exist positive constants $c_3, \tilde{c}_3$ such that for all cubes $Q \subset Q_0$ and every level $\ell > 0$ there holds*

$$\left|\left\{x \in Q : |f(x) - (f)_Q| > \ell\right\}\right| \le c_3 \exp(-\tilde{c}_3 \ell)|Q| \,.$$

(iv) *There exist positive constants $c_4, \tilde{c}_4$ such that for all cubes $Q \subset Q_0$ there holds*

$$\fint_Q \left[\exp(c_4|f - (f)_Q|) - 1\right] dx \le \tilde{c}_4 \,.$$

(v) *There exist positive constants $c_5, \tilde{c}_5$ such that for all cubes $Q \subset Q_0$ there holds*

$$\left(\fint_Q \exp(c_5 f) \, dx\right)\left(\fint_Q \exp(-c_5 f) \, dx\right) \le \tilde{c}_5 \,.$$

Finally, we wish to state and prove the aforementioned characterization of Hölder continuous functions via Campanato spaces.

Theorem 1.27 (Campanato) *Let Ω be a bounded, open set in $\mathbb{R}^n$ which satisfies Ahlfor's regularity condition (1.5) for some $A > 0$. Then, for every $\alpha \in (0, 1]$ and $p \in [1, \infty)$, we have the isomorphy*

$$\mathcal{L}^{p,n+p\alpha}(\Omega, \mathbb{R}^N) \simeq C^{0,\alpha}(\overline{\Omega}, \mathbb{R}^N) \,,$$

and also the semi-norms $[\cdot]_{\mathcal{L}^{p,n+p\alpha}(\Omega,\mathbb{R}^N)}$ and $[\cdot]_{C^{0,\alpha}(\overline{\Omega},\mathbb{R}^N)}$ are equivalent.

Proof The proof is essentially taken from [77, Chapter 1.1, Proof of Lemma 1].
We here show the statement for scalar-valued functions, and the version for
vector-valued functions then follows by considering the single component
functions.

Step 1: $C^{0,\alpha}(\overline{\Omega}) \subset \mathcal{L}^{p,n+p\alpha}(\Omega)$. Let $f \in C^{0,\alpha}(\overline{\Omega})$. By Jensen's inequality
and the definition of Hölder continuity we find for every $x_0 \in \overline{\Omega}$ and $\varrho > 0$:

$$\int_{\Omega(x_0,\varrho)} |f(x) - (f)_{\Omega(x_0,\varrho)}|^p \, dx \leq [f]^p_{C^{0,\alpha}(\overline{\Omega})} |\Omega(x_0,\varrho)| (2\varrho)^{p\alpha}$$

$$\leq c(n,p) [f]^p_{C^{0,\alpha}(\overline{\Omega})} \varrho^{n+p\alpha}.$$

This shows $[f]_{\mathcal{L}^{p,n+p\alpha}(\Omega)} \leq c(n,p)[f]_{C^{0,\alpha}(\overline{\Omega})}$. Furthermore, taking into
account the inequality $\|f\|_{L^p(\Omega)} \leq |\Omega|^{1/p} \|f\|_{C^0(\overline{\Omega})}$, we arrive at

$$\|f\|_{\mathcal{L}^{p,n+p\alpha}(\Omega)} \leq c(n,p,\Omega) \|f\|_{C^{0,\alpha}(\overline{\Omega})},$$

and the claim is proved.

Step 2: Choice of a continuous representative for functions in $\mathcal{L}^{p,n+p\alpha}(\Omega)$.
We start with a preliminary estimate on mean values of a function $f \in$
$\mathcal{L}^{p,n+p\alpha}(\Omega)$ on balls. We take $x_0 \in \overline{\Omega}$ and $0 < r < R \leq \text{diam}(\Omega)$, and we
compute via Hölder's inequality

$$|(f)_{\Omega(x_0,r)} - (f)_{\Omega(x_0,R)}| \leq \left(\fint_{\Omega(x_0,r)} |f(x) - (f)_{\Omega(x_0,R)}|^p \, dx \right)^{\frac{1}{p}}$$

$$\leq |\Omega(x_0,r)|^{-\frac{1}{p}} R^{\frac{n}{p}+\alpha} \left(R^{-n-p\alpha} \int_{\Omega(x_0,R)} |f(x) - (f)_{\Omega(x_0,R)}|^p \, dx \right)^{\frac{1}{p}}$$

$$\leq c(n,p,A) r^{-\frac{n}{p}} R^{\frac{n}{p}+\alpha} [f]_{\mathcal{L}^{p,n+p\alpha}(\Omega)}. \tag{1.6}$$

We now consider the sequence $((f)_{\Omega(x_0,r_j)})_{j\in\mathbb{N}}$ of mean values on domains
$\Omega(x_0,r_j)$ centered at the given point $x_0 \in \overline{\Omega}$ and with radii $r_j = 2^{-j}R$ for
some fixed $0 < R \leq \text{diam}(\Omega)$. Due to the previous inequality (1.6), we have
for $0 \leq j < h$:

$$|(f)_{\Omega(x_0,r_h)} - (f)_{\Omega(x_0,r_j)}| \leq \sum_{\ell=j}^{h-1} |(f)_{\Omega(x_0,r_{\ell+1})} - (f)_{\Omega(x_0,r_\ell)}|$$

$$\leq c(n,p,A) [f]_{\mathcal{L}^{p,n+p\alpha}(\Omega)} R^\alpha \sum_{\ell=j}^{h-1} 2^{(\ell+1)\frac{n}{p}} 2^{-\ell(\frac{n}{p}+\alpha)}$$

$$\leq c(n,p,\alpha,A) [f]_{\mathcal{L}^{p,n+p\alpha}(\Omega)} R^\alpha 2^{-j\alpha}$$

$$= c(n,p,\alpha,A) [f]_{\mathcal{L}^{p,n+p\alpha}(\Omega)} r_j^\alpha, \tag{1.7}$$

and this estimate is *independent* of the point $x_0 \in \overline{\Omega}$. Thus, the sequence of mean values $((f)_{\Omega(x_0,r_j)})_{j \in \mathbb{N}}$ is not only a Cauchy sequence with pointwise limit $f^*(x_0)$ (which due to Lebesgue's differentiation Theorem 1.12 is also a representative of f in $\mathcal{L}^{p,n+p\alpha}(\Omega)$), but even *uniformly convergent*. Moreover, for every fixed radius $r > 0$, the function $x \mapsto (f)_{\Omega(x,r)}$ is continuous. Therefore, f^* is a continuous function (as the uniform limit of a sequence of continuous functions) and precisely the good representative of f we were looking for.

Step 3: Hölder continuity of the continuous representative. Now we take two distinct points $x, y \in \overline{\Omega}$ and set $r := |x - y|$. Then we have

$$|f^*(x) - f^*(y)| \le |f^*(x) - (f)_{\Omega(x,2r)}| + |(f)_{\Omega(x,2r)} - (f)_{\Omega(y,2r)}|$$
$$+ |(f)_{\Omega(y,2r)} - f^*(y)|.$$

Passing to the limit $h \to \infty$ in (1.7), the first and the third term in the latter inequality are estimated by $c(n,p,\alpha,A)[f]_{\mathcal{L}^{p,n+p\alpha}(\Omega)}|x-y|^\alpha$, and we next need to estimate the second term by a similar quantity, in order to bound the α-Hölder semi-norm of f^*. This is done as follows: we first observe the inclusion $\Omega(x,2r) \cap \Omega(y,2r) \supset \Omega(x,r) \cup \Omega(y,r)$ (by recalling the choice $r = |x - y|$). We then calculate via Hölder's inequality and the fact that f belongs to the Campanato space $\mathcal{L}^{p,n+p\alpha}(\Omega)$:

$$|(f)_{\Omega(x,2r)} - (f)_{\Omega(y,2r)}|$$

$$\le \fint_{\Omega(x,2r) \cap \Omega(y,2r)} \left(|(f)_{\Omega(x,2r)} - f(z)| + |f(z) - (f)_{\Omega(y,2r)}|\right) dz$$

$$\le |\Omega(x,r)|^{-1}|\Omega(x,2r)|^{\frac{p-1}{p}} \left(\int_{\Omega(x,2r)} |(f)_{\Omega(x,2r)} - f(z)|^p \, dz\right)^{\frac{1}{p}}$$

$$+ |\Omega(y,r)|^{-1}|\Omega(y,2r)|^{\frac{p-1}{p}} \left(\int_{\Omega(y,2r)} |(f)_{\Omega(y,2r)} - f(z)|^p \, dz\right)^{\frac{1}{p}}$$

$$\le c(n,A)[f]_{\mathcal{L}^{p,n+p\alpha}(\Omega)} r^{-n+n\frac{p-1}{p}+\frac{n+p\alpha}{p}} = c(n,A)[f]_{\mathcal{L}^{p,n+p\alpha}(\Omega)}|x-y|^\alpha.$$

Putting these estimates together, we obtain

$$|f^*(x) - f^*(y)| \le c(n,p,\alpha,A)[f]_{\mathcal{L}^{p,n+p\alpha}(\Omega)}|x-y|^\alpha.$$

Since $x, y \in \overline{\Omega}$ were chosen arbitrarily, we find the following bound for the α-Hölder semi-norm

$$[f^*]_{C^{0,\alpha}(\overline{\Omega})} \le c(n,p,\alpha,A)[f]_{\mathcal{L}^{p,n+p\alpha}(\Omega)}, \tag{1.8}$$

which finishes the assertion on the equivalence of the semi-norms. It only remains to verify that the supremum of f^* is bounded in order to conclude the proof. By the estimate (1.2) from Remark 1.8 – applied with the choice $\ell = \|f\|_{L^p(\Omega)} 2^{1/p} |\Omega(x,1)|^{-1/p}$ – we find for every $x \in \overline{\Omega}$ a subset $\Omega_{\ell,x} \subset \Omega(x,1)$ of measure $|\Omega_{\ell,x}| \geq |\Omega(x,1)|/2 > 0$ such that f is bounded on all of $\Omega_{\ell,x}$ by ℓ (i.e., it is bounded in terms of the L^p-norm of f, p and A). We now pick an arbitrary point $y \in \Omega_{a,y}$. Then, from the previous estimate for the $C^{0,\alpha}$-semi-norm of f^* we get for every $x \in \overline{\Omega}$

$$|f^*(x)| \leq |f^*(x) - f^*(y)| + |f^*(y)|$$
$$\leq c(n,p,\alpha,A)[f]_{\mathcal{L}^{p,n+p\alpha}(\Omega)} + c(p,A)\|f\|_{L^p(\Omega)} . \tag{1.9}$$

In conclusion, we have shown

$$\|f^*\|_{C^{0,\alpha}(\overline{\Omega})} \leq c(n,p,\alpha,A)\|f\|_{\mathcal{L}^{p,n+p\alpha}(\Omega)} ,$$

and hence f possesses a representative in the space $C^{0,\alpha}(\Omega)$ as asserted. This finishes the proof of Campanato's characterization of Hölder continuous functions. □

Remarks 1.28

(i) In similar situations as above, where a continuous representative f^* exists in a Lebesgue (or a Sobolev) class of f, we choose f^* as the representative of its equivalence class and then say that f is continuous.

(ii) Theorem 1.27 holds in particular for domains which have inner cusps or which even have a self-intersecting boundary. In particular, finiteness of the $\mathcal{L}^{p,n+p\alpha}(\Omega, \mathbb{R}^N)$ Campanato norm of a function f, which is defined on one of the latter domains, implies immediately that f can be extended in a unique way up to the boundary.

(iii) The statement of Theorem 1.27 is sharp, in the sense that the isomorphy between the space $\mathcal{L}^{p,n+p\alpha}(\Omega, \mathbb{R}^N)$ and $C^{0,\alpha}(\overline{\Omega}, \mathbb{R}^N)$ does not hold true if Ω is an unbonded domain or if Ahlfor's regularity condition on Ω is violated. However, for $\Omega = \mathbb{R}^n$, the previous proof still yields the continuous embedding $\mathcal{L}^{p,n+p\alpha}(\mathbb{R}^n, \mathbb{R}^N) \hookrightarrow C^{0,\alpha}(\mathbb{R}^n, \mathbb{R}^N)$ with

$$\|f\|_{C^{0,\alpha}(\mathbb{R}^n,\mathbb{R}^N)} \leq c(n,N,p,\alpha)\|f\|_{\mathcal{L}^{p,n+p\alpha}(\mathbb{R}^n,\mathbb{R}^N)} ,$$

while every non-zero constant functions is obviously in $C^{0,\alpha}(\mathbb{R}^n, \mathbb{R}^N)$, but not in $L^p(\mathbb{R}^n, \mathbb{R}^N)$.

(iv) By a rescaling argument, we obtain the explicit dependence of the constant on the domain, for the specific situation of a ball $B_\varrho(x_0)$. In this case we have in particular

$$[g]_{C^{0,\alpha}(\overline{B_\varrho(x_0)},\mathbb{R}^N)} \leq c(n,N,p)[g]_{\mathcal{L}^{p,n+p\alpha}(B_\varrho(x_0),\mathbb{R}^N)}$$

and

$$\|g\|_{C^0(\overline{B_\varrho(x_0)},\mathbb{R}^N)} \le c(n,N,p)\big(\varrho^{-\frac{n}{p}}\|g\|_{L^p(B_\varrho(x_0),\mathbb{R}^N)}$$
$$+ \varrho^\alpha [g]_{\mathcal{L}^{p,n+p\alpha}(B_\varrho(x_0),\mathbb{R}^N)}\big).$$

Proof of (iv) Since balls satisfy Ahlfor's regularity condition, the statement is clear for the unit ball B_1. In the general case, given a function $g\colon B_\varrho(x_0) \to \mathbb{R}^N$ we define the rescaled function $f\colon B_1 \to \mathbb{R}^N$ via $f(y) := g(x_0+\varrho y)$. Then, via the change of variables formula and direct computation, the relevant norms and semi-norms of f and g are related as follows

$$\|g\|_{C^0(\overline{B_\varrho(x_0)},\mathbb{R}^N)} = \|f\|_{C^0(\overline{B_1},\mathbb{R}^N)},$$

$$\|g\|_{L^p(B_\varrho(x_0),\mathbb{R}^N)} = \varrho^{\frac{n}{p}} \|f\|_{L^p(B_1,\mathbb{R}^N)},$$

$$[g]_{C^{0,\alpha}(\overline{B_\varrho(x_0)},\mathbb{R}^N)} = \varrho^{-\alpha} [f]_{C^{0,\alpha}(\overline{B_1},\mathbb{R}^N)},$$

$$[g]_{\mathcal{L}^{p,n+p\alpha}(B_\varrho(x_0),\mathbb{R}^N)} = \varrho^{-\alpha} [f]_{\mathcal{L}^{p,n+p\alpha}(B_1,\mathbb{R}^N)}.$$

Thus, the two claims follow directly from the Campanato isomorphy on the unit ball B_1.   □

1.1.4 Sobolev Spaces

We next discuss the Sobolev spaces $W^{k,p}$ of integer order $k \in \mathbb{N}$ and integrability exponent $p \in [1,\infty]$. For the specific elliptic problems that are in the focus of these lecture notes, the Sobolev spaces turn out to be the natural spaces, in which a concept of a weak solution can be introduced (and its regularity then be studied).

There are two classical ways to give a definition of Sobolev spaces. Either one first introduces the norm (1.10) for smooth functions and then defines the associated Sobolev space by taking the closure with respect to this norm. Or one introduces as a starting point the notion of weak derivatives, which is motivated from the integration by parts formula for differentiable functions, and then defines the Sobolev space as the space of all functions for which both the function and its weak derivatives up to order k belong to the Lebesgue space L^p. We prefer to follow the second approach, because of the similarity to the concept of weak solutions presented later.

Definition 1.29 Let Ω be an open set in $\mathbb{R}^n$, $p \in [1,\infty]$, and let $\beta \in \mathbb{N}_0^n$ be a multiindex. We say that a function $f \in L^1_{\text{loc}}(\Omega,\mathbb{R}^N)$ has a β-th *weak (or distributional) partial derivative in* $L^p_{\text{loc}}(\Omega,\mathbb{R}^N)$ if there exists a function g_β (denoted by $D^\beta f$) in $L^p_{\text{loc}}(\Omega, \mathbb{R}^N)$ such that for every test function $\varphi \in C_0^\infty(\Omega, \mathbb{R}^N)$ we have

$$\int_\Omega f \cdot D^\beta \varphi \, dx = (-1)^{|\beta|} \int_\Omega g_\beta \cdot \varphi \, dx.$$

If for some $k \in \mathbb{N}$ the β-th weak partial derivatives of f exist in $L^1_{\text{loc}}(\Omega, \mathbb{R}^N)$ for all multiindices $\beta \in \mathbb{N}_0^n$ with $0 \le |\beta| \le k$, then we say that f is *weakly differentiable up to order k* (or simply *weakly differentiable* in the case $k = 1$).

Remarks 1.30

(i) If a weak partial derivative exists, then it is uniquely determined up to a set of Lebesgue measure zero. Consequently, the weak derivatives of a differentiable function coincide with the classical ones.

(ii) In contrast to the corresponding classical derivative, the weak derivative $D^\beta f$ is defined globally on all over Ω (but from its definition it is obvious that $D^\beta f$ is the weak derivative on subsets $\Omega' \subset \Omega$, and hence this definition can be localized to a certain extent). As a second distinction between classical and weak derivatives, we note that for the definition of higher order weak derivatives we do not need to assume the existence of the corresponding lower order derivatives (and in fact, this is not necessarily the case, as the example $f(x_1, x_2) = \text{sign}(x_1) + \text{sign}(x_2)$ shows, for which $D_1 D_2 f = 0$ exists on $B_1(0)$, but neither does $D_1 f$ or $D_2 f$).

(iii) A different (but equivalent) way to introduce the concept of weak derivatives is via approximation, in the following sense: the β-th weak partial derivative of a function $f \in L^1_{\text{loc}}(\Omega, \mathbb{R}^N)$ exists in $L^p(\Omega, \mathbb{R}^N)$ if and only if for every open set $O \Subset \Omega$ there exists a sequence of functions $(f_j)_{j \in \mathbb{N}}$ in $C^{|\beta|}(\Omega, \mathbb{R}^N)$ such that f_j converges strongly to f in $L^p(O, \mathbb{R}^N)$ and $(D^\beta f_j)_{j \in \mathbb{N}}$ is a Cauchy sequence in $L^p(O, \mathbb{R}^N)$, see [40, Proposition 3.3].

As anticipated above, the Sobolev spaces $W^{k,p}(\Omega, \mathbb{R}^N)$ of integer order $k \in \mathbb{N}$ are now defined as those subspaces of the $L^p(\Omega, \mathbb{R}^N)$-spaces for which the weak derivatives up to order k exist and are again in $L^p(\Omega, \mathbb{R}^N)$.

Definition 1.31 Let Ω be an open set in $\mathbb{R}^n$, $k \in \mathbb{N}$ and $p \in [1, \infty]$.

(i) We denote by $W^{k,p}(\Omega, \mathbb{R}^N)$ the *Sobolev space* of functions $f \in L^p(\Omega, \mathbb{R}^N)$ such that the weak derivatives $D^\beta f$ exist in $L^p(\Omega, \mathbb{R}^N)$ for all multiindices $\beta \in \mathbb{N}_0^n$ with $0 \le |\beta| \le k$. We endow this space with the norm

$$\|f\|_{W^{k,p}(\Omega,\mathbb{R}^N)} := \begin{cases} \left(\displaystyle\sum_{0 \le |\beta| \le k} \|D^\beta f\|^p_{L^p(\Omega,\mathbb{R}^N)} \right)^{\frac{1}{p}} & \text{if } 1 \le p < \infty \\ \displaystyle\sum_{0 \le |\beta| \le k} \|D^\beta u\|_{L^\infty(\Omega,\mathbb{R}^N)} & \text{if } p = \infty. \end{cases}$$

$$(1.10)$$

(ii) For $p \in [1, \infty)$ we denote by $W_0^{k,p}(\Omega, \mathbb{R}^N)$ the closure of $C_0^\infty(\Omega, \mathbb{R}^N)$ in $W^{k,p}(\Omega, \mathbb{R}^N)$, that is

$$W_0^{k,p}(\Omega, \mathbb{R}^N) := \{ f \in W^{k,p}(\Omega, \mathbb{R}^N) : \text{there exists } (f_j)_{j \in \mathbb{N}} \text{ in } C_0^\infty(\Omega, \mathbb{R}^N)$$
$$\text{with } f_j \to f \text{ in } W^{k,p}(\Omega, \mathbb{R}^N) \}$$

(note that, in view of Mazur's Lemma A.9, we obtain the same space if the strong convergence is replaced by weak convergence in $W^{k,p}(\Omega, \mathbb{R}^N)$, which is equivalent to weak convergence of each derivative up to order k in $L^p(\Omega, \mathbb{R}^N)$). For $p = \infty$ we define $W_0^{k,\infty}(\Omega, \mathbb{R}^N)$ as the sequential closure of $C_0^\infty(\Omega, \mathbb{R}^N)$ in $W^{k,\infty}(\Omega, \mathbb{R}^N)$ with respect to weak-$*$ convergence of each derivative up to order k in $L^\infty(\Omega, \mathbb{R}^N)$.

(iii) For scalar-valued functions (i.e. $N = 1$), we write $W^{k,p}(\Omega)$ instead of $W^{k,p}(\Omega, \mathbb{R})$ and $W_0^{k,p}(\Omega)$ instead of $W_0^{k,p}(\Omega, \mathbb{R})$.

Remarks 1.32

(i) The operator $D^\beta \colon W^{k,p}(\Omega, \mathbb{R}^N) \to W^{k-|\beta|,p}(\Omega, \mathbb{R}^N)$ is linear and continuous.

(ii) An equivalent norm in the Sobolev space $W^{k,p}(\Omega, \mathbb{R}^N)$ is given by $\sum_{0 \le |\beta| \le k} \| D^\beta f \|_{L^p(\Omega, \mathbb{R}^N)}$ for all $p \in [1, \infty]$ and $f \in W^{k,p}(\Omega, \mathbb{R}^N)$.

(iii) Via the application of Hölder's inequality we can verify the inclusions $W^{k,q}(\Omega, \mathbb{R}^N) \subset W^{k,p}(\Omega, \mathbb{R}^N)$ for all $p, q \in [1, \infty]$ with $q \ge p$ provided that Ω is of finite Lebesgue measure.

To a large extent, the theory of the Sobolev spaces $W^{k,p}(\Omega, \mathbb{R}^N)$ is the same (or very similar) as for their Lebesgue counterparts $L^p(\Omega, \mathbb{R}^N)$. In many instances, the properties of the Sobolev spaces are inherited from the Lebesgue spaces (with the additional benefit that one is able to take advantage of the existence of weak derivatives). For illustration, we show the analogous statement to Theorem 1.18, namely that the Sobolev spaces are complete normed spaces, and then only comment on some further properties in Remark 1.35.

Theorem 1.33 *Endowed with the norm* $\| \cdot \|_{W^{k,p}(\Omega, \mathbb{R}^N)}$ *defined in* (1.10), *the Sobolev spaces* $W^{k,p}(\Omega, \mathbb{R}^N)$ *are Banach spaces for all* $p \in [1, \infty]$ *and* $k \in \mathbb{N}$.

Proof
Step 1: (1.10) *defines a norm.* Point separation and absolute homogeneity are again obvious, whereas the triangle inequality is a direct consequence of Minkowski's inequality for L^p-spaces: for $p \in \{1, \infty\}$ this it immediate, and for $p \in (1, \infty)$ it is obtained from the following calculation. Let $f, g \in$

$W^{k,p}(\Omega, \mathbb{R}^N)$. Then we find via Remark 1.19

$$
\begin{aligned}
\|f + g\|_{W^{k,p}(\Omega, \mathbb{R}^N)} &= \Big(\sum_{0 \le |\beta| \le k} \|D^\beta(f+g)\|_{L^p(\Omega, \mathbb{R}^N)}^p \Big)^{\frac{1}{p}} \\
&\le \Big(\sum_{0 \le |\beta| \le k} \big(\|D^\beta f\|_{L^p(\Omega, \mathbb{R}^N)} + \|D^\beta g\|_{L^p(\Omega, \mathbb{R}^N)} \big)^p \Big)^{\frac{1}{p}} \\
&\le \Big(\sum_{0 \le |\beta| \le k} \|D^\beta f\|_{L^p(\Omega, \mathbb{R}^N)}^p \Big)^{\frac{1}{p}} + \Big(\sum_{0 \le |\beta| \le k} \|D^\beta g\|_{L^p(\Omega, \mathbb{R}^N)}^p \Big)^{\frac{1}{p}} \\
&= \|f\|_{W^{k,p}(\Omega, \mathbb{R}^N)} + \|g\|_{W^{k,p}(\Omega, \mathbb{R}^N)}.
\end{aligned}
$$

Step 2: Completeness of $W^{k,p}(\Omega, \mathbb{R}^N)$. Let $(f_j)_{j \in \mathbb{N}}$ be a Cauchy sequence in $W^{k,p}(\Omega, \mathbb{R}^N)$. Then, for each $\beta \in \mathbb{N}_0^n$ with $|\beta| \le k$, the sequence $(D^\beta f_j)_{j \in \mathbb{N}}$ is a Cauchy sequence in $L^p(\Omega, \mathbb{R}^N)$ and therefore, as a consequence of the completeness of the L^p-spaces from Theorem 1.18, it converges to a function $f^\beta \in L^p(\Omega, \mathbb{R}^N)$. In particular, for $\beta = (0, \dots, 0)$, we have the strong convergence $f_j \to f^{(0,\dots,0)} =: f$ in $L^p(\Omega, \mathbb{R}^N)$. To finish the proof, we now verify $f \in W^{k,p}(\Omega, \mathbb{R}^N)$ by showing that its weak derivatives $D^\beta f$ are given by the functions $f^\beta \in L^p(\Omega, \mathbb{R}^N)$, for every multiindex $\beta \in \mathbb{N}_0^n$ with $|\beta| \le k$. To this end we employ the definition of weak derivatives and compute, for every $\varphi \in C_0^\infty(\Omega, \mathbb{R}^N)$, via the strong convergences of both sequences $(f_j)_{j \in \mathbb{N}}$ and $(D^\beta f_j)_{j \in \mathbb{N}}$ combined with the weak differentiability of f_j for each $j \in \mathbb{N}$:

$$
\begin{aligned}
\int_\Omega f \cdot D^\beta \varphi \, dx &= \lim_{j \to \infty} \int_\Omega f_j \cdot D^\beta \varphi \, dx \\
&= \lim_{j \to \infty} (-1)^{|\beta|} \int_\Omega D^\beta f_j \cdot \varphi \, dx = \int_\Omega f^\beta \cdot \varphi \, dx. \qquad \square
\end{aligned}
$$

Remark 1.34 The spaces $W^{k,2}(\Omega, \mathbb{R}^N)$, for $k \in \mathbb{N}$, are Hilbert spaces, with inner product given by

$$
\langle f, g \rangle_{W^{k,2}(\Omega, \mathbb{R}^N)} := \sum_{0 \le |\alpha| \le k} \langle D^\alpha f, D^\alpha g \rangle_{L^2(\Omega, \mathbb{R}^N)}
$$

for all functions $f, g \in W^{k,2}(\Omega, \mathbb{R}^N)$.

Moreover, the Sobolev spaces $W^{k,p}(\Omega, \mathbb{R}^N)$ are reflexive for every $k \in \mathbb{N}$ if and only if $p \in (1, \infty)$. As a consequence of Theorem A.10, bounded sets in $W^{k,p}(\Omega, \mathbb{R}^N)$ are weakly precompact provided that $p \in (1, \infty)$.

Remarks 1.35

(i) *Lebesgue points of Sobolev functions:* We will see in Sect. 1.4 that not only $\mathcal{L}^n$-almost every point of a function in $W^{k,p}(\Omega, \mathbb{R}^N)$ for $k \geq 1$ and $p \in [1, \infty]$ is a Lebesgue point, but in fact every point in Ω outside of a set of *reduced* Hausdorff dimension.

(ii) *Traces of Sobolev functions:* Related to the previous remark, one can also make sense of boundary values (called the trace) of a Sobolev function for sufficiently regular domains. Sobolev functions $W_0^{k,p}(\Omega, \mathbb{R}^N)$ were already introduced as the completion of $C_0^\infty(\Omega, \mathbb{R}^N)$ in $W^{k,p}(\Omega, \mathbb{R}^N)$, so in an abstract sense we have Dirichlet classes $u_0 + W_0^{k,p}(\Omega, \mathbb{R}^N)$ for some $u_0 \in W^{k,p}(\Omega, \mathbb{R}^N)$ available, simply as all equivalence classes f such that $f - u_0$ belongs to $W_0^{k,p}(\Omega, \mathbb{R}^N)$. However, it is also possible to assign to each $f \in W^{k,p}(\Omega, \mathbb{R}^N)$ boundary values in the space $L^p(\partial\Omega)$ (and even better), and this is a bounded, linear operation, which acts as the restriction operator to $\partial\Omega$ whenever f is in addition continuous on $\overline{\Omega}$. For a discussion of this issue we refer to [40, Chapter 3.7].

(iii) *Approximation by smooth functions:* The Sobolev spaces can also be defined via approximation by smooth functions. One way, as mentioned before, is the definition as the closure of $C^\infty(\Omega, \mathbb{R}^N) \cap W^{k,p}(\Omega, \mathbb{R}^N)$ in $W^{k,p}(\Omega, \mathbb{R}^N)$. This definition is equivalent for all $p \in [1, \infty)$ without any further assumption on Ω, see [61], i.e. $C^\infty(\Omega, \mathbb{R}^N) \cap W^{k,p}(\Omega, \mathbb{R}^N)$ is dense in $W^{k,p}(\Omega, \mathbb{R}^N)$. Another way is the definition as the closure of $C^\infty(\overline{\Omega}, \mathbb{R}^N)$ in $W^{k,p}(\Omega, \mathbb{R}^N)$. It turns out that this definition is equivalent for $p \in [1, \infty)$ only under additional assumptions on Ω (such as $\Omega = \mathbb{R}^n$ or Ω regular and bounded; density fails already for the sliced unit ball $\{x \in B_1(0) \subset \mathbb{R}^2 : x_1 > 0 \text{ or } x_2 \neq 0\}$ or more general domains with inner cusps).

The density of $C^\infty(\Omega, \mathbb{R}^N)$ is very useful for proving certain inequalities for $W^{k,p}$-functions which only involve (parts of) the $W^{k,p}$-norms. In such situations it is sufficient to prove these inequalities only for smooth functions (with the advantage that classical derivatives instead of the weak ones may be used), and the general inequality then follows by approximation, see for example the proof of Sobolev's Theorem 1.51.

Finally, we present a statement that concerns the possibility of extending a Sobolev function $f \in W^{k,p}(\Omega)$ outside of Ω, in such a way that on the one hand the extension belongs to $W^{k,p}(\mathbb{R}^n)$ and that on the other hand the $W^{k,p}(\mathbb{R}^n)$-norm of the extension is bounded in terms of the $W^{k,p}(\Omega)$-norm of the original function. It turns out that this operator can only exist if the domain is sufficiently regular, cf. [79, Chapter VI.3].

Theorem 1.36 *Let Ω be a bounded, open set in $\mathbb{R}^n$ with Lipschitz boundary, $k \in \mathbb{N}$ and $p \in [1, \infty]$. For any set $\Omega' \supseteq \Omega$ there exists a bounded, linear extension operator $E \colon W^{k,p}(\Omega, \mathbb{R}^N) \to W^{k,p}(\mathbb{R}^n, \mathbb{R}^N)$ such that we*

have $Ef = f$ *almost everywhere in* Ω *and* $\operatorname{spt} Ef \subset \Omega'$ *for every* $f \in$
$W^{k,p}(\Omega, \mathbb{R}^N)$. *In this case* Ef *is called an* extension *of* f *to* Ω'.

The advantage of the existence of such an extension operator is the
following: one can now prove certain inequalities and embedding theorems
only for functions in $W_0^{1,p}(\Omega, \mathbb{R}^N)$ with zero boundary values, and, under
a suitable regularity assumption on the boundary of Ω, the corresponding
result then follows for all functions in $W^{1,p}(\Omega, \mathbb{R}^N)$ with non-zero boundary
values (see e.g. the proof of Theorems 1.51 and 1.54).

1.1.5   Fractional Sobolev Spaces

Finally, we give the definition of Sobolev spaces of fractional order (also called
Sobolev–Slobodeckij spaces) and comment on a few properties. However, let
us note that there are again several approaches to introduce these extensions
of the classical Sobolev spaces. These are conceptually different and may
therefore lead to (slightly) different families of spaces. Usually, one makes a
choice according to the actual purpose of its introduction, and so we have
decided to follow Gagliardo's approach which was initiated in [28]. For more
details on various fractional order spaces, embeddings between these spaces,
and further properties we refer to [1, Chapter VII] and the recent survey [19].

Definition 1.37 Let Ω be an open set in $\mathbb{R}^n$, let $p \in [1, \infty)$, $k \in \mathbb{N}_0$ and
$\theta \in (0, 1)$.

(i) We denote by $W^{\theta,p}(\Omega, \mathbb{R}^N)$ the set of (equivalence classes of)
    functions $f \in L^p(\Omega, \mathbb{R}^N)$ such that the function $(x, y) \mapsto$
    $|f(x) - f(y)||x - y|^{-n/p-\theta}$ belongs to $L^p(\Omega \times \Omega)$. We endow this space
    with the norm

$$\|f\|_{W^{\theta,p}(\Omega,\mathbb{R}^N)} := \left(\|f\|_{L^p(\Omega,\mathbb{R}^N)} + \int_\Omega \int_\Omega \frac{|f(x) - f(y)|^p}{|x-y|^{n+\theta p}} \, dx \, dy \right)^{\frac{1}{p}} .$$

(ii) For $p \in [1, \infty)$ we denote by $W_0^{\theta,p}(\Omega, \mathbb{R}^N)$ the closure of $C_0^\infty(\Omega, \mathbb{R}^N)$ in
     $W^{\theta,p}(\Omega, \mathbb{R}^N)$.

(iii) The higher fractional Sobolev spaces $W^{k+\theta,p}(\Omega, \mathbb{R}^N)$ are defined accord-
      ingly as the set of all functions $f \in W^{k,p}(\Omega, \mathbb{R}^N)$ such that the weak
      derivative $D^\beta f$ belongs to $W^{\theta,p}(\Omega, \mathbb{R}^N)$ for every $\beta \in \mathbb{N}_0^n$ with $|\beta| = k$.
      We endow this space with the norm

$$\|f\|_{W^{k+\theta,p}(\Omega,\mathbb{R}^N)} := \left(\|f\|_{W^{k-1,p}(\Omega,\mathbb{R}^N)}^p + \sum_{|\beta|=k} \|D^\beta f\|_{W^{\theta,p}(\Omega,\mathbb{R}^N)}^p \right)^{\frac{1}{p}} .$$

$$(1.11)$$

(iv) For scalar-valued functions (i.e. $N = 1$), we write $W^{k+\theta,p}(\Omega)$ instead of $W^{k+\theta,p}(\Omega,\mathbb{R})$ and $W_0^{k+\theta,p}(\Omega)$ instead of $W_0^{k+\theta,p}(\Omega,\mathbb{R})$.

Remarks 1.38

 (i) This definition of fractional differentiability does not distinguish between different directions (and no "partial" fractional derivative is defined).
 (ii) Following a different approach for the definition of Sobolev spaces of fractional order one could obtain similar function spaces via interpolation between the Lebesgue spaces $L^p(\Omega,\mathbb{R}^N)$ and the classical Sobolev spaces $W^{k,p}(\Omega,\mathbb{R}^N)$ for $k \in \mathbb{N}$.
(iii) Another (and slightly different) definition can be given via an approach with pointwise inequalities, which has its origin in the definition of the Calderón spaces in [18]. Here, the Sobolev spaces of fractional order are interpreted as a natural extension of the Hölder spaces, in the sense that a function $f \in L^p(\Omega,\mathbb{R}^N)$ belongs to the fractional Sobolev space $\mathcal{C}^{\theta,p}(\Omega,\mathbb{R}^N)$ if and only if there holds

$$|f(x) - f(y)| \le |x - y|^\theta\big(g(x) + g(y)\big) \qquad \text{for all } x,y \in \Omega_0\,,$$

for some function $g \in L^p(\Omega)$ and some set $\Omega_0 \subset \Omega$ with $\mathcal{L}^n(\Omega \setminus \Omega_0) = 0$. As a norm on this space one can take the sum of the L^p-norm of the function f and the infimum over the L^p-norm of all such functions g. This point of view to regard the fractional Sobolev space as an extension of the space of Hölder continuous functions is supported by the fact that, for domains satisfying Ahlfor's regularity condition (1.5) and for $p \in (1,\infty)$, one can show that $f \in \mathcal{C}^{\theta,p}(\Omega,\mathbb{R}^N)$ is equivalent to

$$\fint_{\Omega(x_0,\varrho)} |f - (f)_{\Omega(x_0,\varrho)}|\, dx \le \varrho^\theta h(x_0)$$

for almost all $x_0 \in \Omega$, small radii ϱ, and some function $h \in L^p(\Omega)$. This statement is proved analogously as Theorem 1.27 on the characterization of Hölder continuous functions via suitable Campanato spaces. Later on, such pointwise estimates (which do not make use of any notion of derivative) were employed for the definition of Sobolev spaces in the context of arbitrary metric spaces, see [44].

Many properties of the Sobolev spaces of fractional order are again inherited from the classical one and from the Lebesgue spaces.

Theorem 1.39 *Endowed with the (Gagliardo) norm* $\|\cdot\|_{W^{k+\theta,p}(\Omega,\mathbb{R}^N)}$ *defined in (1.11), the fractional Sobolev spaces* $W^{k+\theta,p}(\Omega,\mathbb{R}^N)$ *are Banach spaces for all* $\theta \in (0,1)$, $k \in \mathbb{N}_0$, *and every* $p \in [1,\infty)$.

Remarks 1.40

(i) We find the inclusions $W^{\theta',p}(\Omega, \mathbb{R}^N) \subset W^{\theta,p}(\Omega, \mathbb{R}^N)$ whenever $\theta \leq \theta'$ (by distinguishing the cases where $|x - y|$ in the denominator in the $\| \cdot \|_{W^{\theta,p}(\Omega, \mathbb{R}^N)}$-norm is small and large, respectively).

(ii) If Ω has a Lipschitz boundary, the above inclusion continues to hold for the classical Sobolev spaces, i.e. $W^{k+\theta',p}(\Omega, \mathbb{R}^N) \subset W^{k+1,p}(\Omega, \mathbb{R}^N)$ for all $p \in [1, \infty)$, $k \in \mathbb{N}_0$ and $\theta' \in (0, 1)$. However, if Ω is less regular, this inclusion may fail (see [19, Example 9.1]), since the term $|x - y|^{-n-\theta'p}$ in the definition of the fractional Sobolev norm acts as a singular weight and is in general not comparable with the existence of weak derivatives in the L^p-space.

1.2 Criteria for Weak Differentiability

For the Sobolev spaces introduced in the previous section we encounter the problem that – apart from special cases – it might be quite difficult to check explicitly the integration by parts rule from Definition 1.29, to then decide whether or not a given function belongs to a certain Sobolev space. Again, there are several possibilities available, and their applicability depends on the type of functions under investigation. Hence, we now discuss only two criteria that are convenient for later applications and easy to handle (presented here only for scalar-valued functions, which then extends to the vectorial case $N > 1$ by application to the single component functions).

Weak differentiability via classical derivatives on large sets We first discuss a criterion (following the exposition in [80, Chapter 8]) that is suitable in situations in which we already have a natural candidate for the weak derivative, namely in the particular case when the function under consideration is known to have a classical derivative outside of a "small" set.

Lemma 1.41 *Let Ω be a bounded, open set in $\mathbb{R}^n$ and consider $f \in L^q(\Omega) \cap C^1(\Omega \setminus E)$ with $Df \in L^p(\Omega \setminus E, \mathbb{R}^n)$ for some $1 \leq p \leq q \leq \infty$ and some subset $E \subset \Omega$. If this exceptional set E satisfies*

$$\inf \left\{ \|\psi\|_{W^{1,q'}(\mathbb{R}^n, [0,1])} : \psi \in C_0^\infty(\mathbb{R}^n) \text{ with } \psi \geq \mathbb{1}_E \right\} = 0, \qquad (1.12)$$

then we have $f \in W^{1,p}(\Omega)$, and its weak derivative Df coincides almost everywhere with the classical derivative.

Proof We fix a test function $\varphi \in C_0^\infty(\Omega)$ and a coordinate direction $i \in \{1, \ldots, n\}$. By assumption, we can choose a sequence of functions $(\psi_j)_{j \in \mathbb{N}}$ in $C_0^\infty(\mathbb{R}^n, [0, 1])$ such that $\psi_j \geq \mathbb{1}_E$ for all $j \in \mathbb{N}$ and such that, as $j \to \infty$, there hold

$$\|\psi_j\|_{W^{1,q'}(\mathbb{R}^n)} \to 0 \quad \text{and} \quad \psi_j(x) \to 0 \text{ for almost every } x \in \mathbb{R}^n.$$

Then, since the function $\varphi(1 - \psi_j)$ belongs to $C_0^\infty(\Omega \setminus E)$, we can apply the classical integration by parts formula to find

$$\lim_{j \to \infty} \int_\Omega f D_i\big(\varphi(1 - \psi_j)\big) \, dx = - \lim_{j \to \infty} \int_\Omega D_i f \varphi(1 - \psi_j) \, dx = - \int_\Omega D_i f \varphi \, dx \,,$$

where we have also used Lebesgue's dominated convergence theorem (note $|1 - \psi_j| \leq 1$) combined with $D_i f \varphi \in L^p(\Omega)$ (with $D_i f$ extended by 0 to all of Ω). Thus, we get

$$\int_\Omega f D_i \varphi \, dx = \lim_{j \to \infty} \int_\Omega f D_i\big(\varphi(1 - \psi_j) + \varphi \psi_j\big) \, dx$$

$$= - \int_\Omega D_i f \varphi \, dx + \lim_{j \to \infty} \int_\Omega f\big(D_i \varphi \psi_j + \varphi D_i \psi_j\big) \, dx \,.$$

The last term vanishes due to the convergence $\|\psi_j\|_{W^{1,q'}(\mathbb{R}^n)} \to 0$ as $j \to \infty$, because both functions $f\varphi$ and $f D_i \varphi$ belong to $L^q(\Omega)$. Hence, since φ was arbitrary, we have verified that Df satisfies the integration by parts formula, which in turn shows $f \in W^{1,p}(\Omega)$. Therefore, the assertion of the lemma is proved. $\square$

Remark 1.42 The requirement (1.12) on the exceptional set E means that it is of vanishing $W^{1,q'}$-*capacity*. Similarly as the Hausdorff measures, which will be recalled in Sect. 1.4, such capacities are convenient to study lower-dimensional subsets in $\mathbb{R}^n$, see [25, Chapter 4.7]. We further note that it is in general not sufficient to know the classical differentiability outside of a set of Lebesgue measure zero, since the Cantor function is precisely a function which is differentiable in the classical sense almost everywhere, but which has no weak derivative (though it is still in the space BV of functions of bounded variation).

Examples 1.43

(i) *Single points, e.g. $E = \{0\}$, satisfy the condition (1.12) for all $q' \in [1, n]$ and $n \geq 2$.*

(ii) *More generally, the sets $\{x \in B_1(0) : x_1 = \ldots = x_k = 0\} \subset \mathbb{R}^n$ with $k \in \{1, \ldots, n\}$ satisfy (1.12) for all $q' \in [1, k]$.*

(iii) *The function $f(x) := |x|^{-\alpha} x$ with $\alpha \in (0, n)$ belongs to $W^{1,p}(B_1, \mathbb{R}^n)$ for all $p \in [1, n/\alpha)$.*

(iv) More generally, the function $f(x', x'') := |x'|^{-\alpha} x'$ for $(x', x'') \in \mathbb{R}^k \times \mathbb{R}^{n-k}$ with $k \in \{1, \ldots, n\}$ and $\alpha \in (0, k)$ belongs to $W^{1,p}(B_1, \mathbb{R}^n)$ for all $p \in [1, k/\alpha)$.

Proof of (i) and (iii) To verify (i) for the prototypical case $E = \{0\}$ we start by defining a sequence $(\tilde{\psi}_j)_{j \in \mathbb{N}}$ in $W^{1,n}(\mathbb{R}^n, [0, 1])$ via

$$\tilde{\psi}_j(x) := \begin{cases} 1 & \text{if } |x| \leq \exp(-\exp(j+1)), \\ \log(-\log|x|) - j & \text{if } \exp(-\exp(j+1)) < |x| < \exp(-\exp(j)), \\ 0 & \text{if } |x| \geq \exp(-\exp(j)). \end{cases}$$

These functions are rotationally symmetric, satisfy $\psi_j = 1$ in a small neighbourhood of the origin, and they have compact support in a small ball with vanishing radius as $j \to \infty$. It is further straightforward to check

$$\int_{\mathbb{R}^n} |D\tilde{\psi}_j|^n \, dx = c(n) \int_{\exp(-\exp(j+1))}^{\exp(-\exp(j))} |\log r|^{-n} r^{-1} \, dr$$

$$= c(n) \big(\exp(j)^{-n+1} - \exp(j+1)^{-n+1} \big) \to 0 \quad \text{as } j \to \infty,$$

hence we find $\|\tilde{\psi}_j\|_{W^{1,n}(\mathbb{R}^n)} \to 0$. By regularization of the functions $\tilde{\psi}_j$ with suitable (and j-dependent) mollifying kernels, we then obtain a sequence of functions $(\psi_j)_{j \in \mathbb{N}}$ in $C_0^\infty(\mathbb{R}^n, [0, 1])$ satisfying the same properties. At this stage, the validity of condition (1.12) for all $q' \in [1, n]$ is a consequence of Hölder's inequality.

In order to verify the claim (iii), we first observe that f belongs to $L^q(B_1, \mathbb{R}^n)$ with $q = \infty$ for $\alpha \in (0, 1]$ and with every $q < n/(\alpha - 1)$ for $\alpha \in (1, n)$, hence we have in particular $f \in L^{n/(n-1)}(B_1, \mathbb{R}^n)$. The function f further has a classical derivative outside of the origin (hence, according to (i), the exceptional set satisfies (1.12) for $q' = (n/(n-1))' = n$), and

$$D_i f^\kappa = |x|^{-\alpha} \delta_{i\kappa} - \alpha |x|^{-\alpha-2} x_i x_\kappa \quad \in L^p(B_1(0)) \text{ for all } p < \frac{n}{\alpha}$$

for every $i, \kappa \in \{1, \ldots, n\}$. The assertion in (iii) then follows directly from Lemma 1.41. $\qquad \square$

Weak differentiability via L^p-estimates for difference quotients We next provide a criterion (following the presentation of [39, Chapter 7.11]), which is based on the concept of finite difference quotients and which was introduced by Nirenberg [72]. It does not involve weak derivatives and is particularly tailored to functions for which certain integral inequalities involving finite differences can easily be verified.

Definition 1.44 Let Ω be an open set in $\mathbb{R}^n$, $h \in \mathbb{R}$, and $e \in \mathbb{S}^{n-1}$ a unit vector. The *finite difference operator* $\tau_{e,h}$ in direction e and with stepsize h is defined via

$$\tau_{e,h} f(x) \equiv \tau_{e,h}(f)(x) := f(x + he) - f(x)$$

for all $x \in \Omega \cap (\Omega - he)$ and any function $f \colon \Omega \to \mathbb{R}^N$. Accordingly, the *finite difference quotient operator* $\triangle_{e,h}$ in direction e and with stepsize $h \neq 0$ is defined via

$$\triangle_{e,h} f(x) \equiv \triangle_{e,h}(f)(x) := \frac{f(x + he) - f(x)}{h}.$$

Furthermore, if $e = e_s$, $s \in \{1, \ldots, n\}$, is a standard basis vector, then we use the abbreviations $\tau_{s,h}$ and $\triangle_{s,h}$ instead of $\tau_{e_s,h}$ and $\triangle_{e_s,h}$.

Remarks 1.45 From the definition of finite difference quotients we deduce the following elementary properties (with $e \in \mathbb{S}^{n-1}$ and $h \in \mathbb{R} \setminus \{0\}$ as above).

(i) We have a product rule

$$\triangle_{e,h}(fg)(x) = f(x + he)\triangle_{e,h}g(x) + \triangle_{e,h}f(x)g(x)$$

whenever $f, g \colon \Omega \to \mathbb{R}$ and $x \in \Omega \cap (\Omega - he)$.

(ii) The difference quotient operator commutes with weak derivatives, that is, for every $f \colon \Omega \to \mathbb{R}$ and $i \in \{1, \ldots, n\}$ we have $\triangle_{e,h} D_i f = D_i \triangle_{e,h} f$ in $\Omega \cap (\Omega - he)$.

(iii) We have an integration by parts formula for difference quotients:

$$\int_{\mathbb{R}^n} \triangle_{e,h} f g \, dx = -\int_{\mathbb{R}^n} f \triangle_{s,-h} g \, dx$$

for all $f \in L^p(\Omega)$ and $g \in L^{p'}(\Omega)$ (where we interpret the two functions as extended by 0 outside of Ω such that both integrals make sense). In particular, if $f(x) = 0$ for almost every x outside of $\Omega \cap (\Omega + he)$, we can replace the domain of integration in both integrals above by Ω. This is for example the case for $f \in C_0(\Omega)$ and $|h| < \operatorname{dist}(\operatorname{spt} f, \partial\Omega)$.

For a measurable, p-integrable function finite difference quotients are related to weak derivatives via the following two lemmata, which essentially state that uniform (in h) L^p-boundedness of the difference quotients is a necessary and sufficient condition for the existence of a weak derivative in L^p.

Lemma 1.46 *Let Ω be an open set in $\mathbb{R}^n$, $s \in \{1, \ldots, n\}$, and $f \in W^{1,p}(\Omega)$ for some $p \in [1, \infty)$. Then we have $\triangle_{s,h} f \in L^p(\Omega')$ for every subset $\Omega' \Subset \Omega$*

and all $h \in \mathbb{R}$ such that $|h| < \mathrm{dist}(\Omega', \partial\Omega)$, with the estimate

$$\|\triangle_{s,h}f\|_{L^p(\Omega')} \le \|D_sf\|_{L^p(\Omega)}.$$

Proof Let us initially assume $f \in C^1(\Omega) \cap W^{1,p}(\Omega)$ and take Ω' and h as in the statement. Then we have

$$\triangle_{s,h}f(x) = \int_0^1 D_sf(x + hte_s)\,dt$$

for every $x \in \Omega'$. Consequently, by Hölder's inequality and Fubini's theorem we find

$$\int_{\Omega'} |\triangle_{s,h}f(x)|^p\,dx \le \int_{\Omega'} \int_0^1 |D_sf(x + hte_s)|^p\,dt\,dx$$

$$= \int_0^1 \int_{\Omega'} |D_sf(x + hte_s)|^p\,dx\,dt \le \int_\Omega |D_sf(x)|^p\,dx.$$

The same integral estimate holds true for any function $f \in W^{1,p}(\Omega)$ by an approximation argument, see Remark 1.35 (iii). Thus, the assertion of the lemma is proved.   □

Remark 1.47 The requirement $\Omega' + [0,1]he_s := \{x + the_s \colon x \in \Omega', t \in [0,1]\} \subset \Omega$ instead of $\Omega' \Subset \Omega$ with $|h| < \mathrm{dist}(\Omega', \partial\Omega)$ is in fact sufficient.

Lemma 1.48 *Let Ω be an open set in $\mathbb{R}^n$, $s \in \{1,\dots,n\}$, and $f \in L^p(\Omega)$ for some $p \in (1,\infty)$. Suppose that there exists a constant $K \ge 0$ such that $\|\triangle_{s,h}f\|_{L^p(\Omega')} \le K$ for all $h \in \mathbb{R} \setminus \{0\}$ and for every $\Omega' \Subset \Omega$ satisfying $|h| < \mathrm{dist}(\Omega', \partial\Omega)$. Then the weak derivative D_sf exists in Ω and satisfies*

$$\|D_sf\|_{L^p(\Omega)} \le K.$$

Moreover, we have strong convergence $\triangle_{s,h}f \to D_sf$ in $L^p_{\mathrm{loc}}(\Omega)$ as $h \to 0$.

Proof We start by choosing a family of parameters $h \in \mathbb{R} \setminus \{0\}$ with cluster point $\{0\}$ and a family of domains Ω_h with $|h| < \mathrm{dist}(\Omega_h, \partial\Omega)$ and $\Omega_h \supset \{x \in \Omega \colon \mathrm{dist}(x, \partial\Omega) > 2|h|\}$. We then define a family of functions (f_h) in $L^p(\Omega)$ by setting $f_h := \triangle_{s,h}f$ in Ω_h and $f_h = 0$ in $\Omega \setminus \Omega_h$. By assumption of the lemma and by weak precompactness of bounded subsets of $L^p(\Omega)$, we can select a sequence $(h_j)_{j \in \mathbb{N}}$ tending to 0 as $j \to \infty$ and a function $f_s \in L^p(\Omega)$ such that the sequence $(f_{h_j})_{j \in \mathbb{N}}$ converges to f_s weakly in $L^p(\Omega)$. By lower semicontinuity of the norm we further have $\|f_s\|_{L^p(\Omega)} \le K$. To prove the first assertion of the lemma, we now verify $f_s \equiv D_sf$. To this end, we take an arbitrary subset $\Omega' \Subset \Omega$, consider a function $\varphi \in C_0^1(\Omega')$ and elements h_j of the sequence $(h_j)_{j \in \mathbb{N}}$ with $2|h_j| < \mathrm{dist}(\Omega', \partial\Omega)$, which in particular guarantees $\Omega' \subset \Omega_{h_j} \subset \Omega$ for all such j. Then via the weak convergence of the sequence $(\triangle_{s,h_j} f)_{j \in \mathbb{N}}$, the integration by parts formula for finite difference quotients, and the uniform convergence $\triangle_{s,h}\varphi \to D_s\varphi$ on compact sets as $h \to 0$, we have

$$\int_\Omega f_s\varphi\, dx = \lim_{j\to\infty} \int_\Omega \triangle_{s,h_j} f\varphi\, dx$$
$$= -\lim_{j\to\infty} \int_\Omega f\triangle_{s,-h_j}\varphi\, dx = -\int_\Omega f D_s\varphi\, dx.$$

Since $\Omega' \Subset \Omega$ was arbitrary, f is weakly differentiable in direction e_s and f_s coincides with $D_s f$ in Ω. Finally, for every C^1-function f the local strong convergence follows directly from the uniform convergence $\triangle_{s,h} f \to D_s f$ on compact sets as $h \to 0$, while for arbitrary functions f we need the following approximation argument: for $\varepsilon > 0$ we first choose a function $f_\varepsilon \in C^1(\Omega) \cap W^{1,p}(\Omega)$ with $\|D_s f - D_s f_\varepsilon\|_{L^p(\Omega)} \le \varepsilon/3$ (possible by density of smooth functions). Given $\Omega' \Subset \Omega$, we then observe, for h sufficiently small, that the estimate in Lemma 1.46 also ensures $\|\triangle_{s,h} f_\varepsilon - \triangle_{s,h} f\|_{L^p(\Omega')} \le \varepsilon/3$. Consequently, the uniform convergence $\triangle_{s,h} f_\varepsilon \to D_s f_\varepsilon$ on Ω' as $h \to 0$ gives

$$\|D_s f - \triangle_{s,h} f\|_{L^p(\Omega')} \le \|D_s f - D_s f_\varepsilon\|_{L^p(\Omega')} + \|D_s f_\varepsilon - \triangle_{s,h} f_\varepsilon\|_{L^p(\Omega')}$$
$$+ \|\triangle_{s,h} f_\varepsilon - \triangle_{s,h} f\|_{L^p(\Omega')} \le \varepsilon$$

for h sufficiently small, and the proof of the lemma is complete.   □

Remark 1.49 For $p = 1$ the assertion of Lemma 1.48 is in general wrong, i.e., uniform boundedness of difference quotients in L^1 does not guarantee the existence of the weak derivative in L^1 (but still BV-regularity).

Rewriting the assumption of Lemma 1.48, we have found the criterion for weak differentiability that, whenever $\|\tau_{s,h} f\|_{L^p(\Omega')} \le |h| K$ holds for some $s \in \{1,\ldots,n\}$, all h and Ω' as above, then the weak derivative $D_s f$ exists in $L^p(\Omega)$. Concerning fractional Sobolev spaces, we obtain an analogous criterion if $\|\tau_{s,h} f\|_{L^p(\Omega')}$ is uniformly bounded in terms of (noninteger) powers of $|h|$, for all $s \in \{1,\ldots,n\}$, cf. [51, Lemma 2.5].

Lemma 1.50 *Let Ω be an open set in $\mathbb{R}^n$ and consider $f \in L^p(\Omega)$ for some $p \in [1,\infty)$. Suppose that there exist an exponent $\theta \in (0,1]$, a subset $\Omega' \Subset \Omega$ and a constant $K \ge 0$ such that $\|\tau_{s,h} f\|_{L^p(\Omega')} \le |h|^\theta K$ holds for all $s \in \{1,\ldots,n\}$ and every $h \in [-1,1]$ satisfying $|h| < \operatorname{dist}(\Omega',\partial\Omega)$. Then we have $f \in W^{\theta',p}(\Omega'')$ for every $\theta' \in (0,\theta)$ and $\Omega'' \Subset \Omega'$. Moreover, there holds*

$$\int_{\Omega''} \int_{\Omega''} \frac{|f(x) - f(y)|^p}{|x-y|^{n+\theta'p}}\, dx\, dy \le c(n,p) \left(\frac{K^p d^{(\theta-\theta')p}}{\theta - \theta'} + \frac{|\Omega''|}{d^{n+\theta'p}} \int_{\Omega''} |f|^p\, dx \right)$$

for $d := \min\{1, \operatorname{dist}(\Omega',\partial\Omega), \operatorname{dist}(\Omega'',\partial\Omega')\}$.

Proof We may suppose that f is smooth, since the general statement then follows via approximation. For a vector $v = \sum_{s=1}^{n} v_s e_s \in \mathbb{R}^n$ we write $v^{(k)} = \sum_{s=1}^{k} v_s e_s$ for $k = 1, \ldots, n$ and $v^{(0)} = 0$. Then we can decompose the difference $f(x+v) - f(x)$ into differences of f along the coordinate directions and find

$$|f(x+v) - f(x)| = \left| \sum_{s=1}^{n} T_{s,v_s} f(x + v^{(s-1)}) \right| \leq \sum_{s=1}^{n} |T_{s,v_s} f(x + v^{(s-1)})|$$

whenever $x + v^{(s-1)} \in \Omega$ for all $s \in \{0, 1, \ldots, n\}$. We next fix $\Omega'' \Subset \Omega'$ and take $d > 0$ as defined above. Now we employ the assumption of the lemma and calculate

$$\int_{\Omega''} |f(x+v) - f(x)|^p \, dx \leq \int_{\Omega''} \left(\sum_{s=1}^{n} |T_{s,v_s} f(x + v^{(s-1)})| \right)^p dx$$

$$\leq n^{p-1} \int_{\Omega''} \sum_{s=1}^{n} |T_{s,v_s} f(x + v^{(s-1)})|^p \, dx \leq n^p K^p |v|^{p\theta}$$

for all $v \in \mathbb{R}^n$ with $|v| < d$. Hence, we obtain for each $\theta' \in (0, \theta)$:

$$\int_{\{0 < |v| < d\}} \int_{\Omega''} \frac{|f(x+v) - f(x)|^p}{|v|^{n+\theta' p}} \, dx \, dv$$

$$\leq n^p K^p \int_{\{0 < |v| < d\}} |v|^{-n+(\theta - \theta')p} \, dv \leq c(n, p) \frac{K^p d^{(\theta - \theta')p}}{\theta - \theta'}.$$

Taking into account the symmetry with respect to x and y, we thus infer an estimate for all points $(x, y) \in \Omega'' \times \Omega''$ satisfying $|x - y| < d$ (that is, for all points which are close to the diagonal):

$$\int_{\{(x,y) \in \Omega'' \times \Omega'' : |x-y| < d\}} \frac{|f(y) - f(x)|^p}{|y - x|^{n+\theta' p}} \, dx \, dy \leq c(n, p) \frac{K^p d^{(\theta - \theta')p}}{\theta - \theta'}.$$

Otherwise, when considering the remaining points $(x, y) \in \Omega'' \times \Omega''$ satisfying $|x - y| \geq d$, we use the L^p-estimate for f to find

$$\int_{\{(x,y) \in \Omega'' \times \Omega'' : |x-y| \geq d\}} \frac{|f(y) - f(x)|^p}{|y - x|^{n+\theta' p}} \, dx \, dy \leq 2^p d^{-n-\theta' p} |\Omega''| \int_{\Omega''} |f|^p \, dx.$$

Combining the last two inequalities we arrive at the desired estimate. $\square$

1.3   Embedding Theorems and Inequalities

We next discuss some further characteristics of Sobolev spaces, namely (continuous or compact) embedding theorems for the classical Sobolev spaces $W^{1,p}$ into Lebesgue spaces L^q with some $q > p$, or even into Hölder spaces for p sufficiently large. In other words, we will see how differentiability of a function can be used to show its higher integrability or even continuity. We here concentrate on Sobolev spaces of first order, for which the form of embedding depends essentially on the choice of p, more precisely, the theory splits into the three cases $p < n$, $p = n$ and $p > n$. Concerning higher order (and fractional) Sobolev spaces, we note that the corresponding theory is developed analogously and therefore, it will only be stated for completeness. Furthermore, we will also address the relevant Poincaré-type inequalities in this chapter.

All statements in this section depend crucially on the boundary data. It is precisely for this reason that we always suppose that the trace of the function under consideration vanishes, or that the boundary of the domain is at least Lipschitz regular. For the investigation of the relevance of the regularity of the domain and for examples of domains (and arbitrary traces) for which these embeddings fail, we refer to the monograph [59].

Embedding theorems for $p < n$ We start with the famous Sobolev inequality.

Theorem 1.51 (Sobolev, $p \in [1, n)$) *Let Ω be an open set in $\mathbb{R}^n$ and $p \in [1, n)$.*

(i) *The embedding $W_0^{1,p}(\Omega, \mathbb{R}^N) \hookrightarrow L^{np/(n-p)}(\Omega, \mathbb{R}^N)$ is continuous with*

$$\|f\|_{L^{np/(n-p)}(\Omega,\mathbb{R}^N)} \leq c(n, N, p)\|Df\|_{L^p(\Omega,\mathbb{R}^{Nn})}.$$

(ii) *If Ω is bounded and has a Lipschitz-boundary, then the embedding $W^{1,p}(\Omega, \mathbb{R}^N) \hookrightarrow L^{np/(n-p)}(\Omega, \mathbb{R}^N)$ is continuous with*

$$\|f\|_{L^{np/(n-p)}(\Omega,\mathbb{R}^N)} \leq c(n, N, p, \Omega)\|f\|_{W^{1,p}(\Omega,\mathbb{R}^N)}.$$

Proof We here follow the proof of [24, Chapter 5.6, Theorem 1]. We further prove the statement only for scalar-valued functions, which in turn implies a version for vector-valued functions, by considering the component functions.

Step 1: A version for C_0^1-functions and $p = 1$. We start with the proof of the estimate in (i) for functions $f \in C_0^1(\Omega)$ (extended by 0 outside of Ω) and $p = 1$. We first note

$$f(x) = \int_{-\infty}^{x_i} D_i f(x_1, \ldots, x_{i-1}, \xi_i, x_{i+1}, \ldots, x_n) \, d\xi_i$$

$$= -\int_{x_i}^{\infty} D_i f(x_1, \ldots, x_{i-1}, \xi_i, x_{i+1}, \ldots, x_n) \, d\xi_i$$

for every $x \in \mathbb{R}^n$ and each $i \in \{1, \ldots, n\}$. Consequently, we have

$$(2|f(x)|)^{\frac{n}{n-1}} \leq \prod_{i=1}^{n} \left(\int_{\mathbb{R}} |D_i f(x_1, \ldots, x_{i-1}, \xi_i, x_{i+1}, \ldots, x_n)| \, d\xi_i \right)^{\frac{1}{n-1}}. \quad (1.13)$$

We next want to integrate this inequality with respect to x. To this end we prove by induction that we have

$$\int_{\mathbb{R}^\ell} (2|f(x)|)^{\frac{n}{n-1}} \, dx_1 \ldots dx_\ell \qquad\qquad\qquad (1.14)$$

$$\leq \prod_{i=1}^{\ell} \left(\int_{\mathbb{R}^\ell} |D_i f(x)| \, dx_1 \ldots dx_\ell \right)^{\frac{1}{n-1}}$$

$$\times \prod_{i=\ell+1}^{n} \left(\int_{\mathbb{R}^{\ell+1}} |D_i f(x_1, \ldots, x_{i-1}, \xi_i, x_{i+1}, \ldots, x_n)| \, dx_1 \ldots dx_\ell \, d\xi_i \right)^{\frac{1}{n-1}}$$

for all $\ell \in \{1, \ldots, n\}$. To verify this inequality for $\ell = 1$, we first integrate (1.13) with respect to x_1 (noting that the first factor on the right-hand side of (1.13) is independent of x_1), then we apply the generalized Hölder inequality from Corollary 1.17 (to the remaining $n - 1$ factors with all exponents equal to $n - 1$) and finally Fubini's theorem. This gives

$$\int_{\mathbb{R}} (2|f(x)|)^{\frac{n}{n-1}} \, dx_1 \leq \left(\int_{\mathbb{R}} |D_1 f(\xi_1, x_2, \ldots, x_n|) \, d\xi_1 \right)^{\frac{1}{n-1}}$$

$$\times \int_{\mathbb{R}} \prod_{i=2}^{n} \left(\int_{\mathbb{R}} |D_i f(x_1, \ldots, \xi_i, \ldots, x_n)| \, d\xi_i \right)^{\frac{1}{n-1}} dx_1$$

$$\leq \left(\int_{\mathbb{R}} |D_1 f(x_1, x_2, \ldots, x_n|) \, dx_1 \right)^{\frac{1}{n-1}}$$

$$\times \prod_{i=2}^{n} \int_{\mathbb{R}^2} |D_i f(x_1, \ldots, \xi_i, \ldots, x_n)| \, dx_1 \, d\xi_i \right)^{\frac{1}{n-1}},$$

hence the claim (1.14) is true for $\ell = 1$. Now we assume (1.14) to be true for all $i \leq \ell - 1$ and some $\ell \in \{2, \ldots, n-1\}$. We then integrate $(1.14)_{\ell-1}$ with respect to x_ℓ. Now, the ℓ-th factor on the right-hand side of (1.14) is independent of x_ℓ. Therefore, applying Hölder's inequality and Fubini's theorem in the same way as before, we end up with $(1.14)_\ell$ and hence, we have proved (1.14) for all $\ell \in \{1, \ldots, n\}$. In particular, with (1.14) for $\ell = n$ (and using the generalized inequality between arithmetic and geometric mean), we arrive at the desired inequality in (i) for the particular case $p = 1$:

$$2\|f\|_{L^{n/(n-1)}(\Omega)} = \left(\int_{\mathbb{R}^n} (2|f(x)|)^{\frac{n}{n-1}} \, dx \right)^{\frac{n-1}{n}}$$

$$\leq \prod_{i=1}^{n} \left(\int_{\mathbb{R}^n} |D_i f(x)| \, dx \right)^{\frac{1}{n}}$$

$$\leq n^{-1} \sum_{i=1}^{n} \int_{\mathbb{R}^n} |D_i f(x)| \, dx$$

$$\leq n^{-\frac{1}{2}} \int_{\mathbb{R}^n} |Df(x)| \, dx = n^{-\frac{1}{2}} \|Df\|_{L^1(\Omega, \mathbb{R}^n)}.$$

Step 2: A version for C_0^1-functions and $p > 1$. Next we derive the estimate in (i) for functions $f \in C_0^1(\Omega)$ and arbitrary $p \in (1, n)$, by tracing it back to the estimate with $p = 1$, applied to the function $g := |f|^\gamma$ for some suitable exponent $\gamma > 1$. Since $Dg = \gamma \operatorname{sign}(f)|f|^{\gamma-1} Df$ by the classical chain rule, we find, in view of Hölder's inequality, the estimate

$$\left(\int_{\mathbb{R}^n} |f|^{\frac{n\gamma}{n-1}} \, dx \right)^{\frac{n-1}{n}} \leq c(n, \gamma) \int_{\mathbb{R}^n} |f|^{\gamma-1} |Df| \, dx$$

$$\leq c(n, \gamma) \left(\int_{\mathbb{R}^n} |Df|^p \, dx \right)^{\frac{1}{p}} \left(\int_{\mathbb{R}^n} |f|^{\frac{(\gamma-1)p}{p-1}} \, dx \right)^{\frac{p-1}{p}}.$$

For the specific choice $\gamma = p(n-1)/(n-p) > 1$ (recall that $1 < p < n$ is true by assumption) the exponents $n\gamma/(n-1)$ on the left-hand side and $(\gamma-1)p/$

$(p-1)$ on the right-hand side coincide and are equal to $np/(n-p)$. Hence, we deduce

$$\|f\|_{L^{np/(n-p)}(\Omega)} = \left(\int_{\mathbb{R}^n} |f|^{\frac{np}{n-p}}\, dx\right)^{\frac{n-1}{n}-\frac{p-1}{p}}$$

$$\leq c(n,p)\left(\int_{\mathbb{R}^n} |Df|^p\, dx\right)^{\frac{1}{p}} = c(n,p)\|Df\|_{L^p(\Omega,\mathbb{R}^n)}.$$

Step 3: General estimate in (i) *via approximation.* By Definition 1.31 of the space $W_0^{1,p}(\Omega)$ the space of smooth functions $C_0^\infty(\Omega)$ is dense in $W_0^{1,p}(\Omega)$ with respect to the $W^{1,p}(\Omega)$-norm. Therefore, given an arbitrary function $f \in W_0^{1,p}(\Omega)$ we find an approximating sequence $(f_j)_{j\in\mathbb{N}}$ of functions in $C_0^\infty(\Omega)$ such that $\|f - f_j\|_{W^{1,p}(\Omega)} \to 0$ as $j \to \infty$. Then, involving Step 1 and Step 2 above, we get

$$\|f_j\|_{L^{np/(n-p)}(\Omega)} \leq c(n,p)\|Df_j\|_{L^p(\Omega,\mathbb{R}^n)},$$

$$\|f_j - f_m\|_{L^{np/(n-p)}(\Omega)} \leq c(n,p)\|D(f_j - f_m)\|_{L^p(\Omega,\mathbb{R}^n)}.$$

The second inequality implies that $(f_j)_{j\in\mathbb{N}}$ is a Cauchy sequence in the space $L^{np/(n-p)}(\Omega)$, and thus, it converges to a function $g \in L^{np/(n-p)}(\Omega)$. By uniqueness of the limits, we have $g \equiv f$. Taking the limit in the first of the above inequalities, we have established claim (i).

Step 4: Assertion (ii) *via extension.* Since Ω is assumed to be a bounded set with Lipschitz-continuous boundary, there exists, according to Theorem 1.36, an extension operator $E_\delta \colon W^{1,p}(\Omega) \to W_0^{1,p}(\Omega_\delta)$ with $\Omega_\delta := \{x \in \mathbb{R}^n \colon \operatorname{dist}(x,\Omega) < \delta\}$, for each $\delta > 0$, such that

$$\|E_\delta f\|_{W^{1,p}(\Omega_\delta)} \leq c(n,p,\Omega,\delta)\|f\|_{W^{1,p}(\Omega)}.$$

With the estimate in (i) applied to the function $E_\delta f$ on Ω_δ, we then find

$$\|f\|_{L^{np/(n-p)}(\Omega)} \leq \|E_\delta f\|_{L^{np/(n-p)}(\Omega_\delta)}$$

$$\leq c(n,p)\|D(E_\delta f)\|_{L^p(\Omega_\delta,\mathbb{R}^n)} \leq c(n,p,\Omega,\delta)\|f\|_{W^{1,p}(\Omega)}.$$

This finishes the proof of the Sobolev inequality.

$\square$

Remarks 1.52

(i) The exponent $np/(n-p)$ is called the *Sobolev exponent* to p and is usually abbreviated by p^*. Note that $p^* > p$ with $1/p - 1/p^* = 1/n$.

(ii) For $p = n$ the embedding $W^{1,n}(\Omega) \hookrightarrow L^{\infty}(\Omega)$ does not hold. A counterexample is the function $f(x) = \log\log|x|^{-1}$ with $\Omega = B_{1/e}(0)$. However, in this limiting case (still supposing that Ω is Lipschitz) we have that the embedding $W^{1,n}(\Omega) \hookrightarrow \mathcal{L}^{1,n}(\Omega) = BMO(\Omega)$ is continuous, and the corresponding estimate is referred to as the Moser–Trudinger inequality, cf. [68, 82]. The precise statement is that there exists a constant $\alpha(n)$ such that for every $f \in W_0^{1,n}(\Omega)$ there holds

$$\int_{\Omega} \exp\left(\alpha\frac{|f|}{\|f\|_{W_0^{1,n}(\Omega)}}\right)^{\frac{n}{n-1}} dx \le c(n, \Omega).$$

(iii) There is an interesting connection between the Sobolev inequality and the isoperimetric inequality. The classical isoperimetric inequality states that the closed unit ball in $\mathbb{R}^n$ has the least perimeter among all closed sets $S \subset \mathbb{R}^n$ of the same $\mathcal{L}^n$-measure. If S is regular (in the sense that ∂S is rectifiable), the inequality is given by

$$|S|^{1-\frac{1}{n}} \le c(n)\mathcal{H}^{n-1}(\partial S),$$

where $\mathcal{H}^{n-1}$ is the $(n-1)$-dimensional Hausdorff measure (see Definition 1.68 below) and where the constant $c(n)$ is determined by equality in the case $S = B_1$. By means of the coarea function one can show that the Sobolev inequality for $p = 1$ is equivalent to the isoperimetric inequality, see the discussion in [40, Chapter 3].

Corollary 1.53 *Let Ω be an open set in $\mathbb{R}^n$, $k \in \mathbb{N}$, and $p \in [1, n/k)$.*

(i) *The embedding $W_0^{k,p}(\Omega, \mathbb{R}^N) \hookrightarrow L^{np/(n-kp)}(\Omega, \mathbb{R}^N)$ is continuous with*

$$\|f\|_{L^{np/(n-kp)}(\Omega,\mathbb{R}^N)} \le c(n, N, p, k)\|D^k f\|_{L^p(\Omega,\mathbb{R}^{Nn^k})}.$$

(ii) *If Ω is bounded and has a Lipschitz-boundary, then the embedding $W^{k,p}(\Omega, \mathbb{R}^N) \hookrightarrow L^{np/(n-kp)}(\Omega, \mathbb{R}^N)$ is continuous with*

$$\|f\|_{L^{np/(n-kp)}(\Omega,\mathbb{R}^N)} \le c(n, N, p, \Omega)\|f\|_{W^{k,p}(\Omega,\mathbb{R}^{Nn})}.$$

Proof Applying k-times Theorem 1.51 we find

$$W_0^{k,p}(\Omega, \mathbb{R}^N) \hookrightarrow W_0^{k-1, \frac{np}{n-p}}(\Omega, \mathbb{R}^N)$$

$$\hookrightarrow W_0^{k-2, \frac{np}{n-2p}}(\Omega, \mathbb{R}^N) \hookrightarrow \ldots \hookrightarrow L^{np/(n-kp)}(\Omega, \mathbb{R}^N)$$

with the corresponding estimates for the norms. The second embedding then follows by extension, see again Theorem 1.36.   □

In Theorem 1.51 we have proved that the embedding $W^{1,p}(\Omega, \mathbb{R}^N) \hookrightarrow L^{p^*}(\Omega, \mathbb{R}^N)$ is continuous, when Ω is bounded and regular. Hence, also $W^{1,p}(\Omega, \mathbb{R}^N) \hookrightarrow L^q(\Omega, \mathbb{R}^N)$ is continuous for all exponents $q \in [1, p^*)$. In fact, these embeddings are even compact in the sense of Definition A.4.

Theorem 1.54 (Rellich–Kondrachov, $p \in [1, n)$) *Let Ω be a bounded, open set in $\mathbb{R}^n$ and $p \in [1, n)$.*

(i) *The embedding $W_0^{1,p}(\Omega, \mathbb{R}^N) \hookrightarrow L^q(\Omega, \mathbb{R}^N)$ is compact for every $q \in [1, p^*)$.*

(ii) *If Ω has a Lipschitz boundary, then the embedding $W^{1,p}(\Omega, \mathbb{R}^N) \hookrightarrow L^q(\Omega, \mathbb{R}^N)$ is compact for every $q \in [1, p^*)$.*

Remarks 1.55

(i) The embedding $W^{1,p}(\Omega, \mathbb{R}^N) \hookrightarrow L^{p^*}(\Omega, \mathbb{R}^N)$ is in general not compact, see [1, Example 6.11].

(ii) The Rellich–Kondrachov embedding theorem is an example of the more general principle that bounded sequences in a space of higher regularity with support in a compact domain are often compact in a space of lower regularity. Another example for this phenomenon is the corresponding compactness result for Hölder spaces, which states that the embedding $C^{0,\alpha_1}(\overline{\Omega}, \mathbb{R}^N) \hookrightarrow C^{0,\alpha_2}(\overline{\Omega}, \mathbb{R}^N)$ is compact for all $0 < \alpha_2 < \alpha_1 \le 1$ if $\Omega \subset \mathbb{R}^n$ is bounded.

Proof of Remark 1.55 (ii) We first observe that the embedding $C^{0,\alpha_1}(\overline{\Omega}) \hookrightarrow C^{0,\alpha_2}(\overline{\Omega})$ is continuous, due to Remark 1.5 (iv). Now let $(f_j)_{j \in \mathbb{N}}$ be a bounded sequence in $C^{0,\alpha_1}(\overline{\Omega})$ with $\|f_j\|_{C^{0,\alpha_1}(\overline{\Omega})} \le C_0$ for all $j \in \mathbb{N}$ and some constant C_0. In particular, all functions f_j are uniformly bounded on a compact set, and by assumption they are also equicontinuous. Hence, by Theorem 1.6 of Arzelà–Ascoli, the sequence is relatively compact in $C(\overline{\Omega})$, i.e., there exists a subsequence $(f_{j(\ell)})_{\ell \in \mathbb{N}}$ that converges uniformly to some function $f \in C^{0,\alpha_1}(\overline{\Omega})$, and without loss of generality we may assume $f \equiv 0$. Moreover, this subsequence converges to 0 also in $C^{0,\alpha_2}(\overline{\Omega})$, since we have

$$\frac{|f_{j(\ell)}(x) - f_{j(\ell)}(y)|}{|x-y|^{\alpha_2}} = \left(\frac{|f_{j(\ell)}(x) - f_{j(\ell)}(y)|}{|x-y|^{\alpha_1}} \right)^{\frac{\alpha_2}{\alpha_1}} |f_{j(\ell)}(x) - f_{j(\ell)}(y)|^{1 - \frac{\alpha_2}{\alpha_1}}$$

$$\le 2 C_0^{\frac{\alpha_2}{\alpha_1}} \|f_{j(\ell)}\|_{C^0(\overline{\Omega})}^{1 - \frac{\alpha_2}{\alpha_1}} \to 0 \qquad \text{as } \ell \to \infty,$$

uniformly for all $x, y \in \overline{\Omega}$.

$\square$

Proof of Theorem 1.54 We proceed analogously to the proof of [24, Chapter 5.7, Theorem 1] and restrict ourselves again to scalar-valued functions. We first observe that (ii) follows from (i) in combination with the existence of the extension operator for bounded domains with Lipschitz boundary, in a similar way as in Step 4 of the proof of Theorem 1.51. Moreover, since we have already shown the continuity of the embedding $W_0^{1,p}(\Omega) \hookrightarrow L^{p^*}(\Omega)$, it only remains to verify that, given a bounded sequence $(f_j)_{j \in \mathbb{N}}$ in $W_0^{1,p}(\Omega)$, we find a subsequence, which converges in $L^q(\Omega)$ for all $q \in [1, p^*)$. Hence, in view of Theorem 1.51 (i), we may work under the permanent assumption

$$\|f_j\|_{L^{p^*}(\Omega)} + \|f_j\|_{W^{1,1}(\Omega)} + \|f_j\|_{W^{1,p}(\Omega)} \leq C_0 \qquad \text{for all } j \in \mathbb{N}.$$

Moreover, extending each function f_j by zero outside of Ω, we may regard $(f_j)_{j \in \mathbb{N}}$ as a bounded sequence in $W_0^{1,p}(\mathbb{R}^n)$, and we can then define the mollifications

$$f_j * \eta_\varepsilon(x) := \int_{\mathbb{R}^n} f_j(x - y) \eta_\varepsilon(y) \, dy$$

for $\varepsilon \in (0, 1)$, where the functions η_ε are standard ε-mollifying kernels given by $\eta_\varepsilon(y) := \varepsilon^{-n} \eta(y/\varepsilon)$ for $y \in \mathbb{R}^n$, for a fixed non-negative, rotationally symmetric function $\eta \in C_0^\infty(B_1(0))$ normalized to $\int_{\mathbb{R}^n} \eta \, dx = 1$. In what follows, we are now going to exploit some properties of the family $(f_j * \eta_\varepsilon)_{j \in \mathbb{N}, \varepsilon \in (0,1)}$.

Step 1: For fixed $\varepsilon \in (0, 1)$ the functions $(f_j * \eta_\varepsilon)_{j \in \mathbb{N}}$ are uniformly bounded and equicontinuous. We observe that, for each $x \in \mathbb{R}^n$ and $j \in \mathbb{N}$, we have

$$|f_j * \eta_\varepsilon(x)| \leq \|f_j\|_{L^1(\mathbb{R}^n)} \|\eta_\varepsilon\|_{L^\infty(\mathbb{R}^n)} \leq C_0 C(\eta) \varepsilon^{-n}$$

and

$$|D(f_j * \eta_\varepsilon)(x)| = |f_j * D\eta_\varepsilon(x)|$$
$$\leq \|f_j\|_{L^1(\mathbb{R}^n)} \|D\eta_\varepsilon\|_{L^\infty(\mathbb{R}^n, \mathbb{R}^n)} \leq C_0 C(\eta) \varepsilon^{-n-1}.$$

Step 2: The family $(f_j * \eta_\varepsilon)_{\varepsilon \in (0,1)}$ converges to f_j in $L^q(\mathbb{R}^n)$ as $\varepsilon \searrow 0$, uniformly for $j \in \mathbb{N}$ and for all $q \in [1, p^*)$. The claim for $q = 1$ is established by the following explicit estimate relying on Fubini's theorem (this is done rigorously via approximation, cp. proof of Lemma 1.46):

$$\|f_j * \eta_\varepsilon - f_j\|_{L^1(\mathbb{R}^n)} \leq \int_{\mathbb{R}^n} \int_{\mathbb{R}^n} |f_j(x - y) - f_j(x)| \eta_\varepsilon(y) \, dy \, dx$$
$$\leq \int_{\mathbb{R}^n} \int_{\mathbb{R}^n} \int_0^1 |Df_j(x - ty)| \, dt \, |y| \eta_\varepsilon(y) \, dy \, dx \leq C_0 C(\eta) \varepsilon$$

for all $j \in \mathbb{N}$ and $\varepsilon \in (0, 1)$. The claim for general $q \in (1, p^*)$ then follows in turn from the interpolation inequality in Remark 1.16 (ii) and the fact that mollifications preserve norms, which imply

$$\|f_j * \eta_\varepsilon - f_j\|_{L^q(\mathbb{R}^n)} \leq \|f_j * \eta_\varepsilon - f_j\|_{L^1(\mathbb{R}^n)}^{\theta} \|f_j * \eta_\varepsilon - f_j\|_{L^{p^*}(\mathbb{R}^n)}^{1-\theta} \leq C_0 C(\eta) \varepsilon^{\theta},$$

where $\theta \in (0, 1)$ is chosen such that $1/q = \theta + (1 - \theta)/p^*$ holds.

Step 3: Conclusion. We start by showing that for every $i \in \mathbb{N}$ there exists a subsequence $(f_{j(\ell)})_{\ell \in \mathbb{N}}$ which satisfies

$$\limsup_{\ell, m \to \infty} \|f_{j(\ell)} - f_{j(m)}\|_{L^q(\mathbb{R}^n)} \leq \frac{1}{i}. \tag{1.15}$$

To this end we first select $\varepsilon(i) > 0$ according to Step 2 such that $\|f_j * \eta_\varepsilon - f_j\|_{L^q(\mathbb{R}^n)} \leq (2i)^{-1}$ holds for all $j \in \mathbb{N}$. Then we employ Step 1 and Theorem 1.6 of Arzelà–Ascoli, to choose for this $\varepsilon(i)$ a uniformly convergent subsequence $(f_{j(\ell)} * \eta_{\varepsilon(i)})_{\ell \in \mathbb{N}}$, which then in particular satisfies

$$\limsup_{\ell, m \to \infty} \|f_{j(\ell)} * \eta_{\varepsilon(i)} - f_{j(m)} * \eta_{\varepsilon(i)}\|_{L^q(\mathbb{R}^n)} = 0.$$

The combination of these two facts yields the initial assertion (1.15) for the sequence $(f_{j(\ell)})_{\ell \in \mathbb{N}}$. The existence of subsequence which is a Cauchy sequence in $L^q(\mathbb{R}^n)$ and hence convergent then follows from a diagonal argument applied for all $i \in \mathbb{N}$.

□

Poincaré-type inequalities Next we discuss an inequality that is similar to the previous Sobolev inequality, but with the difference that only the Lebesgue norm of the derivatives, instead of the full Sobolev norm, is involved as upper bound for a suitable Lebesgue norm of the function itself. For $W_0^{1,p}$-functions with integrability exponent $p < n$ this is actually already contained in Theorem 1.51, but we now derive a version for the whole range $p \in [1, \infty)$. Moreover, we prove such inequalities also for Sobolev functions $f \in W^{1,p}$ with possibly non-zero boundary values, imposing only the additional assumptions that Ω is connected and that the mean value of f over this domain vanishes. Without an assumption of this type such inequalities are in general wrong (since we might just add an arbitrary constant).

Lemma 1.56 (Classical Poincaré inequality) *Let Ω be a bounded, open set in $\mathbb{R}^n$ and let $p \in [1, \infty)$.*

(i) *For every function $f \in W_0^{1,p}(\Omega, \mathbb{R}^N)$ we have*

$$\|f\|_{L^p(\Omega, \mathbb{R}^N)} \leq c(n, N, p, \Omega) \|Df\|_{L^p(\Omega, \mathbb{R}^{Nn})}.$$

(ii) *If Ω is connected with Lipschitz-boundary, then for every function $f \in W^{1,p}(\Omega, \mathbb{R}^N)$ we have*

$$\|f - (f)_\Omega\|_{L^p(\Omega, \mathbb{R}^N)} \leq c(n, N, p, \Omega)\|Df\|_{L^p(\Omega, \mathbb{R}^{Nn})}.$$

Proof The inequality in (i) is a direct consequence of Hölder's inequality and Theorem 1.51 (i), applied with p if $p < n$ and with $\frac{np}{n+p} \in [1, n)$ otherwise.

In order to derive the inequality in (ii), we may assume without loss of generality $(f)_\Omega = 0$ since the claimed inequality is invariant under the addition of constants to f. We then argue by contradiction and assume that the lemma were false. Then there would exist a sequence of functions $(f_j)_{j \in \mathbb{N}}$ in $W^{1,p}(\Omega, \mathbb{R}^N)$ with $(f_j)_\Omega = 0$ and $\|f_j\|_{L^p(\Omega, \mathbb{R}^N)} = 1$ (this is achieved by normalization) for all $j \in \mathbb{N}$ such that

$$\|Df_j\|_{L^p(\Omega, \mathbb{R}^{Nn})} \leq j^{-1} \tag{1.16}$$

holds. Hence, the sequence $(f_j)_{j \in \mathbb{N}}$ is bounded in $W^{1,p}(\Omega, \mathbb{R}^N)$, and consequently, due to the Rellich–Kondrachov compactness Theorem 1.54 (which is applied with p if $p < n$ or any number in $(np/(n+p), n)$ if $p \geq n$), there exists a subsequence $(f_{j(\ell)})_{\ell \in \mathbb{N}}$ and a function $f \in W^{1,p}(\Omega, \mathbb{R}^N)$ such that

$$f_{j(\ell)} \to f \quad \text{strongly in } L^p(\Omega, \mathbb{R}^N) \text{ as } \ell \to \infty.$$

The strong convergence implies in particular that the properties $(f)_\Omega = 0$ and $\|f\|_{L^p(\Omega, \mathbb{R}^N)} = 1$ are preserved for the limit. Moreover, in view of (1.16), we find $\|Df\|_{L^p(\Omega, \mathbb{R}^{Nn})} = 0$ (by lower semicontinuity of the norm or by Fatou's lemma; alternatively, one can show $Df = 0$ almost everywhere via the definition of weak derivative, invoking the strong convergence of the subsequence $(f_{j(\ell)})_{\ell \in \mathbb{N}}$). However, since Ω is connected, this is a contradiction to the fact that the limit function f satisfies both identities $\|f\|_{L^p(\Omega, \mathbb{R}^N)} = 1$ and $(f)_\Omega = 0$. $\qquad \square$

Remarks 1.57

(i) For convex domains it is not difficult to give a direct proof of Poincaré's inequality, with similar arguments as in the proof of Lemma 1.66.

(ii) If Ω is a bounded domain with Lipschitz-boundary as in Lemma 1.56 (ii) and if a function $f \in W^{1,p}(\Omega, \mathbb{R}^N)$ vanishes on a subset of Ω of positive measure, i.e. $|\Omega_0| := |\{x \in \Omega : f(x) = 0\}| = \gamma|\Omega|$ for some $\gamma \in (0, 1]$, then we get

$$\|f\|_{L^p(\Omega, \mathbb{R}^N)} \leq c(n, N, p, \Omega, \gamma)\|Df\|_{L^p(\Omega, \mathbb{R}^{Nn})}.$$

This is seen easily by adding $-(f)_\Omega + (f)_\Omega - (f)_{\Omega_0}$ to f, then estimating $(f)_\Omega - (f)_{\Omega_0}$ in terms of $\|f - (f)_\Omega\|_{L^p(\Omega, \mathbb{R}^N)}$ and γ, and finally applying Poincaré's inequality in the mean value version given in Lemma 1.56 (ii).

(iii) The combination of Sobolev's embedding stated in Theorem 1.51 (ii) and of Poincaré's Lemma 1.56 (ii) yields an improved inequality, commonly known as *Sobolev–Poincaré inequality*,

$$\|f - (f)_\Omega\|_{L^{p^*}(\Omega,\mathbb{R}^N)} \leq c(n,N,p,\Omega)\|Df\|_{L^p(\Omega,\mathbb{R}^{Nn})}$$

for all $f \in W^{1,p}(\Omega,\mathbb{R}^N)$ with $p < n$, under the assumptions of the previous Lemma 1.56.

(iv) By the following scaling argument one gets the explicit dependence on Ω for simple but useful domains, such as balls, cp. Remark 1.28 (iv). For $g \in W^{1,p}(B_\varrho(x_0),\mathbb{R}^N)$ the scaled function $f(y) = g(x_0 + \varrho y)$ belongs to $W^{1,p}(B_1(0),\mathbb{R}^N)$. Then, via Poincaré's inequality for f on the unit ball B_1, we get for the constant appearing in Poincaré's inequality for g the correct scaling behavior in terms of the radius ϱ:

$$\|g - (g)_{B_\varrho(x_0)}\|_{L^p(B_\varrho(x_0),\mathbb{R}^N)} = \varrho^{\frac{n}{p}}\|f - (f)_{B_1(0)}\|_{L^p(B_1(0),\mathbb{R}^N)}$$

$$\leq c(n,N,p)\varrho^{\frac{n}{p}}\|Df\|_{L^p(B_1(0),\mathbb{R}^{Nn})}$$

$$= c(n,N,p)\varrho\|Dg\|_{L^p(B_\varrho(x_0),\mathbb{R}^{Nn})}.$$

Accordingly, the constant in the Sobolev–Poincaré inequality from (iii) becomes in this situation ϱ-independent, i.e., we have

$$\|g - (g)_{B_\varrho(x_0)}\|_{L^{p^*}(B_\varrho(x_0),\mathbb{R}^N)} \leq c(n,N,p)\|Dg\|_{L^p(B_\varrho(x_0),\mathbb{R}^{Nn})}.$$

Embedding theorems for $p > n$ The second important embedding theorem concerns the Sobolev space $W^{1,p}(\Omega,\mathbb{R}^N)$ in the particular case that the integrability exponent p is strictly greater than the space dimension n. In this situation the embedding is not only continuous into the space $L^\infty(\Omega,\mathbb{R}^N)$, but even into a suitable Hölder space. Before stating this embedding theorem, we first address a consequence of Campanato's characterization of Hölder continuous functions from Theorem 1.27 for functions in the Sobolev space $W^{1,p}_0(\Omega,\mathbb{R}^N)$, cp. [60].

Corollary 1.58 *Let Ω be an open set in $\mathbb{R}^n$, $\alpha \in (0,1]$, and $p \in [1,\infty)$ with $p(1-\alpha) \leq n$. Then $f \in W^{1,p}_0(\Omega,\mathbb{R}^N)$ with $Df \in L^{p,n-p(1-\alpha)}(\Omega,\mathbb{R}^{Nn})$ implies $f \in C^{0,\alpha}(\overline{\Omega},\mathbb{R}^N)$, with*

$$\|f\|_{C^{0,\alpha}(\overline{\Omega},\mathbb{R}^N)} \leq c(n,N,p,\alpha)\big(\|Df\|_{L^{p,n-p(1-\alpha)}(\Omega,\mathbb{R}^{Nn})} + \|f\|_{L^p(\Omega,\mathbb{R}^N)}\big).$$

Proof Due to the zero-boundary assumption on f, we can extend f outside of Ω by zero and therefore, we may suppose $f \in W^{1,p}_0(\mathbb{R}^n,\mathbb{R}^N)$ with $Df \in L^{p,n-p(1-\alpha)}(\mathbb{R}^n,\mathbb{R}^{Nn})$. From Poincaré's inequality in Lemma 1.56, applied for $x_0 \in \overline{\Omega}$ and $\varrho \leq 1$, and the scaling in the radius according to Remark 1.57 (iv), we thus find

$$\int_{B_\varrho(x_0)} |f - (f)_{B_\varrho(x_0)}|^p \, dx \le c(n, N, p)\varrho^p \int_{B_\varrho(x_0)} |Df|^p \, dx$$

$$\le c(n, N, p)\|Df\|^p_{L^{p, n-p(1-\alpha)}(\Omega, \mathbb{R}^{Nn})} \varrho^{n+p\alpha} \,.$$

Therefore, we arrive at the bound

$$\|f\|_{\mathcal{L}^{p, n+p\alpha}(\mathbb{R}^n, \mathbb{R}^N)} \le c(n, N, p)\big(\|Df\|_{L^{p, n-p(1-\alpha)}(\Omega, \mathbb{R}^{Nn})} + \|f\|_{L^p(\Omega, \mathbb{R}^N)}\big)$$

for every such f. At this point, the assertion with the claimed dependency of the constant follows from Theorem 1.27 and Remark 1.28 (iii).   □

Remark 1.59 Since Hölder continuity does not necessarily imply weak differentiability (cf. Weierstrass functions or Blancmange curves extended to more than one space dimension), we cannot expect equivalence in the statement of the corollary.

Remark 1.60 In a similar way, one obtains also a version of Corollary 1.58 for bounded Lipschitz domains, without the zero-boundary condition. In particular, for balls $B_\varrho(x_0) \subset \mathbb{R}^n$, every function $f \in W^{1,p}(B_\varrho(x_0), \mathbb{R}^N)$ with $Df \in L^{p, n-p(1-\alpha)}(B_\varrho(x_0), \mathbb{R}^{Nn})$ belongs to $f \in C^{0,\alpha}(\overline{B_\varrho(x_0)}, \mathbb{R}^N)$, with the corresponding estimate.

We the previous result at hand, we can now give the embedding theorem for Sobolev functions in $W^{1,p}(\Omega, \mathbb{R}^N)$, for the case $p > n$.

Theorem 1.61 (Morrey's inequality, $p > n$) *Let Ω be an open set in $\mathbb{R}^n$ and let $p \in (n, \infty)$.*

(i) *The embedding $W_0^{1,p}(\Omega, \mathbb{R}^N) \hookrightarrow C^{0,1-n/p}(\overline{\Omega}, \mathbb{R}^N)$ is continuous with*

$$\|f\|_{C^{0,1-n/p}(\overline{\Omega}, \mathbb{R}^N)} \le c(n, N, p)\|f\|_{W^{1,p}(\Omega, \mathbb{R}^N)} \,.$$

(ii) *If Ω is bounded and has a Lipschitz-boundary, then the embedding $W^{1,p}(\Omega, \mathbb{R}^N) \hookrightarrow C^{0,1-n/p}(\overline{\Omega}, \mathbb{R}^N)$ is continuous with*

$$\|f\|_{C^{0,1-n/p}(\overline{\Omega}, \mathbb{R}^N)} \le c(n, N, p, \Omega)\|f\|_{W^{1,p}(\Omega, \mathbb{R}^N)} \,.$$

Proof We start with the proof of (i). Since every function $f \in W_0^{1,p}(\Omega, \mathbb{R}^N)$ trivially satisfies $Df \in L^{p,0}(\Omega, \mathbb{R}^{Nn})$, the claim is a direct consequence of Corollary 1.58, applied with $\alpha = 1 - n/p$.

We next deduce (ii) from (i). Since Ω is a bounded set with Lipschitz-continuous boundary, there exists, according to Theorem 1.36, an extension operator $E_\delta \colon W^{1,p}(\Omega) \to W_0^{1,p}(\Omega_\delta)$ with $\Omega_\delta := \{x \in \mathbb{R}^n \colon \text{dist}(x, \Omega) < \delta\}$ for every $\delta > 0$, which preserves the $W^{1,p}$-norm with a constant $c(n, p, \Omega, \delta)$.

With statement (i) applied to the function $E_\delta f$ on Ω_δ, we then find

$$\|f\|_{C^{0,1-n/p}(\overline{\Omega},\mathbf{R}^N)} \le \|E_\delta f\|_{C^{0,1-n/p}(\overline{\Omega},\mathbf{R}^N)}$$

$$\le c(n,N,p)\|E_\delta f\|_{W^{1,p}(\Omega_\delta,\mathbf{R}^N)}$$

$$\le c(n,p,\Omega,\delta)\|f\|_{W^{1,p}(\Omega,\mathbf{R}^N)}.$$

This finishes the proof of the Morrey's inequality. □

Remark 1.62 In view of Morrey's inequality from Theorem 1.61 and Sobolev's inequality from Theorem 1.51, we obtain in particular that every function $f \in W^{1,1}(\Omega,\mathbf{R}^N)$ with $Df \in L^p(\Omega,\mathbf{R}^{Nn})$ belongs to $W^{1,p}(\Omega,\mathbf{R}^N)$, provided that Ω is a bounded, open set in $\mathbf{R}^n$ with Lipschitz boundary.

Before iterating Sobolev's and Morrey's inequality in order to deduce an embedding result for higher order Sobolev spaces into Hölder spaces, we observe that the concepts of classical and weak derivatives are compatible.

Corollary 1.63 *Let Ω be an open set in $\mathbf{R}^n$ and consider $f \in W^{1,p}(\Omega,\mathbf{R}^N)$ for some $p \in (n,\infty)$. Then f is differentiable in the classical sense in every p-Lebesgue point of Df (and hence, by Corollary 1.13 almost everywhere).*

Proof We follow [54, Proof of Corollary 11.36]. We take a p-Lebesgue point $x_0 \in \Omega$ of Du and a ball $B_\varrho(x_0) \Subset \Omega$. Due to Theorem 1.61, f is continuous, and since the Hölder semi-norm remains invariant under addition of constants, we infer with the help of Poincaré's inequality from Lemma 1.56 the estimate

$$[f]_{C^{0,1-n/p}(\overline{B_\varrho(x_0)},\mathbf{R}^N)} \le c(n,N,p)\|Df\|_{L^p(B_\varrho(x_0),\mathbf{R}^{Nn})}.$$

We here have in addition taken into account that the constant does not depend on the radius r (this is easily seen by a scaling argument). Applying this inequality to the function $g \in W^{1,p}(B_\varrho(x_0),\mathbf{R}^N)$ defined via $g(x) := f(x) - f(x_0) - Df(x_0) \cdot (x - x_0)$, we find

$$|g(x) - g(x_0)| \le c(n,N,p)\varrho^{1-\frac{n}{p}}\|Dg\|_{L^p(B_\varrho(x_0),\mathbf{R}^{Nn})}$$

for all $x \in \partial B_\varrho(x_0)$. Since this inequality is equivalent to

$$\frac{|f(x) - f(x_0) - Df(x_0) \cdot (x - x_0)|}{|x - x_0|}$$

$$\le c(n,N,p)\Big(\fint_{B_\varrho(x_0)} |Df(x) - Df(x_0)|^p \, dx \Big)^{\frac{1}{p}},$$

the claim follows from the definition of p-Lebesgue point (see Corollary 1.13), which ensures that the right-hand side of the latter inequality vanishes in the limit $\varrho \searrow 0$. □

Corollary 1.64 *Let Ω be an open set in $\mathbb{R}^n$, $k, m \in \mathbb{N}$ with $m \leq n$, and $p \in (n/m, n/(m-1))$.*

(i) *The embedding $W_0^{k,p}(\Omega, \mathbb{R}^N) \hookrightarrow C^{k-m,m-n/p}(\overline{\Omega}, \mathbb{R}^N)$ is continuous with*

$$\|f\|_{C^{k-m,m-n/p}(\overline{\Omega},\mathbb{R}^N)} \leq c(n, N, p, k)\|f\|_{W^{k,p}(\Omega,\mathbb{R}^N)}.$$

(ii) *If Ω is bounded and has a Lipschitz boundary, then the embedding $W^{k,p}(\Omega, \mathbb{R}^N) \hookrightarrow C^{k-m,m-n/p}(\Omega, \mathbb{R}^N)$ is continuous with*

$$\|f\|_{C^{k-m,m-n/p}(\Omega,\mathbb{R}^N)} \leq c(n, N, p, k, \Omega)\|f\|_{W^{k,p}(\Omega,\mathbb{R}^N)}.$$

Proof Due to $(m-1)p < n$ we first apply Corollary 1.53, and then, in view of the second inequality $mp > n \Leftrightarrow np/(n-(m-1)p) > n$, Theorem 1.61. In combination with Corollary 1.63, which allows to identify the weak derivative with the classical one, this yields

$$W_0^{k,p}(\Omega, \mathbb{R}^N) \hookrightarrow W_0^{k-m+1,np/(n-(m-1)p)}(\Omega, \mathbb{R}^N) \hookrightarrow C^{k-m,m-n/p}(\overline{\Omega}, \mathbb{R}^N)$$

with the corresponding estimates for the norms. The second embedding then follows by extension, see Theorem 1.36. □

Analogously to the embeddings for the Sobolev spaces $W^{1,p}(\Omega, \mathbb{R}^N)$ with $p < n$ which embeds continuously into $L^{p^*}(\Omega, \mathbb{R}^N)$ and compactly into $L^q(\Omega, \mathbb{R}^N)$ for all $q \in [1, p^*)$, we obtain for the Sobolev spaces $W^{1,p}(\Omega, \mathbb{R}^N)$ with $p > n$ the compact embedding into the Hölder spaces $C^{0,\alpha}(\overline{\Omega}, \mathbb{R}^N)$ for all Hölder exponents $\alpha \in (0, 1 - n/p)$.

Theorem 1.65 (Rellich–Kondrachov, $p > n$) *Let Ω be a bounded, open set in $\mathbb{R}^n$ and let $p \in (n, \infty)$.*

(i) *The embedding $W_0^{1,p}(\Omega, \mathbb{R}^N) \hookrightarrow C^{0,\alpha}(\overline{\Omega}, \mathbb{R}^N)$ is compact for every $\alpha \in (0, 1 - n/p)$.*
(ii) *If Ω has a Lipschitz boundary, then the embedding $W^{1,p}(\Omega, \mathbb{R}^N) \hookrightarrow C^{0,\alpha}(\overline{\Omega}, \mathbb{R}^N)$ is compact for every $\alpha \in (0, 1 - n/p)$.*

Proof With the assumptions on Ω and p, the embeddings $W_0^{1,p}(\Omega, \mathbb{R}^N) \hookrightarrow C^{0,1-n/p}(\overline{\Omega}, \mathbb{R}^N)$ and $W^{1,p}(\Omega, \mathbb{R}^N) \hookrightarrow C^{0,1-n/p}(\overline{\Omega}, \mathbb{R}^N)$ are continuous according to Morrey's inequality from Theorem 1.61. Moreover, the embedding $C^{0,1-n/p}(\overline{\Omega}, \mathbb{R}^N) \hookrightarrow C^{0,\alpha}(\overline{\Omega}, \mathbb{R}^N)$ is compact for every $\alpha \in (0, 1-n/p)$, due to Remark 1.55 (ii). Thus, the embeddings $W_0^{1,p}(\Omega, \mathbb{R}^N) \hookrightarrow C^{0,\alpha}(\overline{\Omega}, \mathbb{R}^N)$ and $W^{1,p}(\Omega, \mathbb{R}^N) \hookrightarrow C^{0,\alpha}(\overline{\Omega}, \mathbb{R}^N)$ are in fact compact, as compositions of a continuous and a compact embedding, see Remark A.5. □

Corresponding results for fractional Sobolev spaces Also for the fractional Sobolev spaces we can give suitable versions of the previous embedding theorems as well as Poincaré-type inequalities. These extend the classical results for the Sobolev spaces $W^{m,p}$ with integer values of m to the fractional ones, and for sake of completeness we include the main statements here. We start by stating a fractional Poincaré inequality for the particular situation of a ball (see e.g. [63, inequality (4.2)]).

Lemma 1.66 (Fractional Poincaré inequality on balls) *Let $B_\varrho(x_0) \subset \mathbb{R}^n$, $p \in [1,\infty)$, and $\theta \in (0,1)$. Then for every function $f \in W^{\theta,p}(B_\varrho(x_0), \mathbb{R}^N)$ we have*

$$\int_{B_\varrho(x_0)} |f - (f)_{B_\varrho(x_0)}|^p \, dx$$

$$\leq c(n,N,p)\varrho^{\theta p} \int_{B_\varrho(x_0)} \int_{B_\varrho(x_0)} \frac{|f(x) - f(y)|^p}{|x-y|^{n+\theta p}} \, dx \, dy \, .$$

Proof Without loss of generality we may assume $x_0 = 0$ and $\varrho = 1$, otherwise one uses a rescaling argument, compare Remark 1.57 (iv). We may further assume $f \in C^0(B_1, \mathbb{R}^N)$, since the general statement then follows by approximation. By Jensen's inequality, we then observe

$$|f(x) - (f)_{B_1}|^p \leq \fint_{B_1} |f(x) - f(y)|^p \, dy$$

$$\leq c(n,p) \fint_{B_1} \frac{|f(x) - f(y)|^p}{\max\{|x-y|^{n+\theta p}, \varepsilon\}} \, dy$$

for every $x \in B_1$ and every $\varepsilon \in (0,1)$. Integration with respect to x yields

$$\int_{B_1} |f(x) - (f)_{B_1}|^p \, dx \leq c(n,p) \int_{B_1} \int_{B_1} \frac{|f(x) - f(y)|^p}{\max\{|x-y|^{n+\theta p}, \varepsilon\}} \, dy \, dx \, .$$

With the passage $\varepsilon \searrow 0$, the claim follows from Theorem 1.10 on monotone convergence. $\square$

Concerning the embedding theory for fractional Sobolev spaces $W^{s,p}(\Omega, \mathbb{R}^N)$, one could give various results and consider also continuous embeddings in other fractional Sobolev spaces, see e.g. [1, Theorem 7.57 and Theorem 7.58]. Restricting ourselves only to embeddings into Lebesgue and Hölder spaces and to fractional Sobolev spaces with order of differentiability $s \leq 1$, these results amount to the following statement.

Theorem 1.67 *Let Ω be a bounded domain in $\mathbb{R}^n$ with Lipschitz boundary. Furthermore, let $s \in (0,1]$, $p \in (1,\infty)$ and assume $f \in W^{s,p}(\Omega, \mathbb{R}^N)$. Then the following statements are true:*

(i) *If $n > sp$, then $f \in L^t(\Omega, \mathbb{R}^N)$ for all $t \leq np/(n - sp)$.*
(ii) *If $n = sp$, then $f \in L^t(\Omega, \mathbb{R}^N)$ for all $t < \infty$.*
(iii) *If $n < sp$, then $f \in C(\overline{\Omega}, \mathbb{R}^N)$.*

1.4   Fine Properties of Sobolev Functions

We next discuss some fine properties of Sobolev functions, with the aim to obtain a better interpretation of Sobolev functions. As we have already discussed before, every Lebesgue function is an equivalence class of measurable functions, and as a consequence of the Lebesgue differentiation Theorem 1.12, it coincides almost everywhere with its Lebesgue representative (defined at every point as the limit of averages on balls centered at this point as the radius goes to zero when this limit exists, and zero otherwise). The requirements on a Sobolev function $f \in W^{1,p}$ that all weak derivatives of first order exist and that they are p-integrable, are additional regularity properties of f, which have also consequences on the "size" of the set of Lebesgue points. For example, Morrey's Theorem 1.61 implies that, for $p > n$, each point is in fact a Lebesgue point of f, and not only $\mathcal{L}^n$-almost everyone. The aim in this section is to determine the size of the set of non-Lebesgue points. Natural ways to do so are via the concept of Hausdorff dimension or via the concept of $W^{1,p}$-capacities (which in turn implies an estimate on the Hausdorff dimension, cp. [25, Chapter 4.7]). Both approaches are classical (the latter was actually already employed for the Sobolev spaces of integer order in the 1950s in [17]), but since we do not want to go into the details of capacities, we will here follow the first approach, which also seems to be the more common one in modern elliptic regularity theory.

Hausdorff measure and Hausdorff dimension We first recall the definition of the k-dimensional Hausdorff measure $\mathcal{H}^k$, which is one possibility to measure the volume of very small sets in $\mathbb{R}^n$ (in fact, "k-dimensional" sets with $k < n$ which are negligible with respect to the $\mathcal{L}^n$-measure). For this purpose, we first introduce the normalizing constants $\omega_k = \pi^{k/2}/\Gamma(1 + k/2)$ (where Γ denotes the Euler Γ-function) for $k \geq 0$. Note that this constant coincides with the volume of the unit ball in $\mathbb{R}^k$ if $k \geq 1$ is an integer (and with this normalization one can in fact prove that the Hausdorff measure $\mathcal{H}^n$ and the Lebesgue measure $\mathcal{L}^n$ coincide on $\mathbb{R}^n$).

Definition 1.68 (Hausdorff measure)   Consider a set $S \subset \mathbb{R}^n$ and a number $k \in [0, \infty)$. The *k-dimensional Hausdorff pre-measure of fineness* $\delta > 0$ of S is defined as

$$\mathcal{H}_\delta^k(S) := \inf \left\{ \sum_{j=1}^{\infty} \omega_k \left(\frac{\operatorname{diam}(S_j)}{2} \right)^k : S \subset \bigcup_{j=1}^{\infty} S_j, \operatorname{diam}(S_j) < \delta \right\}$$

with the convention $\operatorname{diam}(\emptyset) = 0$. The *k-dimensional Hausdorff measure* of S is then defined as

$$\mathcal{H}^k(S) := \lim_{\delta \searrow 0} \mathcal{H}_\delta^k(S).$$

Remarks 1.69

(i) The k-dimensional Hausdorff pre-measure $\mathcal{H}_\delta^k(S)$ of a set $S \subset \mathbb{R}^n$ is monotonically non-increasing in δ since all coverings which are allowed to determine the Hausdorff pre-measure $\mathcal{H}_\delta^k(S)$ of fineness δ are contained in the possible coverings to determine the Hausdorff pre-measure $\mathcal{H}_{\delta'}^k(S)$ of every fineness $\delta' > \delta$. Thus, the limit $\mathcal{H}_\delta^k(S)$ exists (but may be infinite).

(ii) Also special coverings consisting of only balls might be considered. This approach defines the *spherical Hausdorff measure* which is in general strictly larger than $\mathcal{H}^k$.

Proposition 1.70 (Properties of the Hausdorff measure)   *Let S be a set in $\mathbb{R}^n$ and $k \in [0, \infty)$. The Hausdorff measure $\mathcal{H}^k$ in $\mathbb{R}^n$ has the following properties:*

(i)   *$\mathcal{H}^k$ is an outer measure and Borel regular (but not a Radon measure for $k \in [0, n)$ since in this case $\mathcal{H}^k$ is not finite on any ball of positive radius);*

(ii) *$\mathcal{H}^k$ is invariant under isometries $T : \mathbb{R}^n \to \mathbb{R}^n$, i.e. $\mathcal{H}^k(TS) = \mathcal{H}^k(S)$, and it is homogeneous of degree k, i.e. $\mathcal{H}^k(rS) = r^k \mathcal{H}^k(S)$ for all $r \geq 0$;*

(iii) *For $k' \in [0, k)$ and $\mathcal{H}^k(S) > 0$ we have $\mathcal{H}^{k'}(S) = \infty$ (reversely, $\mathcal{H}^{k'}(S) < \infty$ implies $\mathcal{H}^k(S) = 0$); moreover, $\mathcal{H}^k$ is identically zero for $k > n$;*

(iv) *If $f : \mathbb{R}^n \to \mathbb{R}^m$ is Lipschitz continuous, then $\mathcal{H}^k(f(S)) \leq [f]_{C^{0,1}}^k \mathcal{H}^k(S)$.*

In view of property (iii) we can assign to every set $S \subset \mathbb{R}^n$ a unique number d with the property that $\mathcal{H}^k(S) = 0$ for every $k > d$ and $\mathcal{H}^{k'}(S) = \infty$ for every $k' < d$. This justifies the following definition of Hausdorff dimension.

Definition 1.71 (Hausdorff dimension)   The *Hausdorff dimension* of a set S in $\mathbb{R}^n$ is defined as

$$\dim_{\mathcal{H}}(S) := \inf \left\{ k \geq 0 : \mathcal{H}^k(S) = 0 \right\}.$$

By monotonicity of the Hausdorff measure, we obviously have $\dim_{\mathcal{H}}(S') \leq \dim_{\mathcal{H}}(S)$ whenever S' and S are subsets of $\mathbb{R}^n$ with $S' \subset S$.

A measure density result We next give a covering lemma and a measure density result which traces its origins back to Giusti and which will play the central role in order to control the Hausdorff dimension of the set of the non-Lebesgue points of Sobolev functions.

Lemma 1.72 (Vitali covering lemma) *Let $\mathcal{G}$ be an arbitrary family of closed balls B in $\mathbb{R}^n$ with radius $r(B) \in (0, R]$ for some uniform constant $R < \infty$. There exists an at most countable subfamily $\mathcal{G}'$ of pairwise disjoint balls such that*

$$\bigcup_{B \in \mathcal{G}} B \subset \bigcup_{B \in \mathcal{G}'} \widehat{B} \quad \text{with } \widehat{B} = B_{5r}(x_0) \text{ if } B = B_r(x_0).$$

Proof For every $j \in \mathbb{N}$ we define

$$\mathcal{G}_j := \left\{ B \in \mathcal{G} : r(B) \in (2^{-j}R, 2^{-j+1}R] \right\}.$$

We choose $\mathcal{G}'_1$ as a maximal subfamily of pairwise disjoint balls of $\mathcal{G}_j$ (due to $r(B) > R/2$ for all $B \in \mathcal{G}_1$ this subfamily is at most countable). The choice of this maximal subfamily is possible by the Hausdorff maximal principle (which in turn is a consequence of the axiom of choice). Now we assume that the subfamilies $\mathcal{G}'_1, \ldots, \mathcal{G}'_m$ are already defined for $m \in \mathbb{N}$. Then we choose $\mathcal{G}'_{m+1}$ as a maximal subfamily of pairwise disjoint balls in the family

$$\left\{ B \in \mathcal{G}_{m+1} : B \cap B' = \emptyset \text{ for all } B' \in \bigcup_{j \leq m} \mathcal{G}'_j \right\}.$$

Again, the subfamily $\mathcal{G}'_{m+1}$ is at most countable. By construction, the subfamily $\mathcal{G}' = \cup_{j \in \mathbb{N}} \mathcal{G}'_j$ consists of pairwise disjoint balls and is at most countable (as union of countably many families of at most countably many elements). Moreover, given a ball $B \in \mathcal{G}$, we have $B \in \mathcal{G}_m$ for some $m \in \mathbb{N}$, and by maximality of the subfamilies there exists some $B' \in \cup_{j \leq m} \mathcal{G}'_j$ with $B \cap B' \neq \emptyset$. Since $2r(B) \geq r(B')$, the claim follows. $\qquad \square$

Remark 1.73 The number 5 in the statement is not optimal and can be replaced by any number greater than 3 (but not 3).

Now we can state the aforementioned measure density result. We give the statement from [63, Section 4], but we note that a basic version of this result (for a positive Radon measure λ) was formulated and proved by Giusti in [40, Proposition 2.7].

Lemma 1.74 (Giusti) *Let Ω be an open set in $\mathbb{R}^n$, and let λ be a finite, non-negative and non-decreasing function which is defined on the family of open subsets of Ω and which is countably super-additive in the sense that*

$$\sum_{i\in\mathbb{N}}\lambda(O_i) \leq \lambda\Big(\bigcup_{i\in\mathbb{N}}O_i\Big)$$

holds whenever $\{O_i\}_{i\in\mathbb{N}}$ is a family of pairwise disjoint open subsets of Ω. Then, for every $\alpha \in (0,n)$, we have $\dim_{\mathcal{H}}(E^\alpha) \leq \alpha$ where

$$E^\alpha := \big\{x_0 \in \Omega: \limsup_{\varrho\searrow 0} \varrho^{-\alpha}\lambda\big(\Omega(x_0,\varrho)\big) > 0\big\}.$$

Proof We decompose E^α into the sets

$$E_j := \big\{x_0 \in \Omega: \limsup_{\varrho\searrow 0} \varrho^{-\alpha}\lambda\big(\Omega(x_0,\varrho)\big) > j^{-1}\big\}$$

with $j \in \mathbb{N}$. Since this decomposition is countable, it is sufficient to prove that $\mathcal{H}^{\alpha+\varepsilon}(E_j) = 0$ holds for every fixed $j \in \mathbb{N}$ and $\varepsilon > 0$. We now choose $\delta > 0$ as a parameter for the fineness of the Hausdorff pre-measure. By definition of E_j we infer that for every $x \in E_j$ there exists a ball $B_{\varrho(x)}(x) \subset \Omega$ with radius $\varrho(x) < \delta$ such that

$$\lambda(B_{\varrho(x)}(x)) > \varrho^\alpha j^{-1},$$

and obviously, the inclusion $E_j \subset \cup_{x\in E_j} B_{\varrho(x)}(x)$ holds. By Vitali's covering Lemma 1.72 we find an at most countable subfamily of pairwise disjoint balls $(B_{\varrho_i}(x_i))_{i\in I}$ (with $\varrho_i := \varrho(x_i)$) such that

$$E_j \subset \bigcup_{i\in I} B_{5\varrho_i}(x_i).$$

Thus, by the countable super-additivity of λ, we have

$$\mathcal{H}_{5\delta}^{\alpha+\varepsilon}(E_j) \leq \sum_{i\in I}\omega_{\alpha+\varepsilon}(5\varrho_i)^{\alpha+\varepsilon}$$

$$\leq \omega_{\alpha+\varepsilon}\delta^\varepsilon 5^{\alpha+\varepsilon}\sum_{i\in I}\varrho_i^\alpha$$

$$\leq \omega_{\alpha+\varepsilon}\delta^\varepsilon 5^{\alpha+\varepsilon} j\sum_{i\in I}\lambda(B_{\varrho_i}(x_i))$$

$$\leq \omega_{\alpha+\varepsilon}\delta^\varepsilon 5^{\alpha+\varepsilon} j\lambda\big(\{x \in \Omega: \operatorname{dist}(x, E_j) < \delta\}\big) \leq \omega_{\alpha+\varepsilon}\delta^\varepsilon 5^{\alpha+\varepsilon} j\lambda(\Omega).$$

Letting $\delta \searrow 0$, we arrive at $\mathcal{H}^{\alpha+\varepsilon}(E_j) = 0$ for all $j \in \mathbb{N}$ and every $\varepsilon > 0$, and the proof of the lemma is complete. □

Remark 1.75 Under additional assumptions on λ it is also possible to obtain the even stronger conclusion $\mathcal{H}^{\alpha}(E^{\alpha}) = 0$ with essentially the same line of arguments (and $\varepsilon = 0$). For instance, one might be interested in the classical case (considered by Giusti) that λ is defined via integration with respect to a non-negative function $g \in L^1(\mathbb{R}^n)$, that is

$$\lambda(O) := \int_O g(x)\, dx \qquad \text{for every open set } O \subset \mathbb{R}^n \,.$$

In this case, λ is even additive on disjoint open sets, and moreover, in view of the integrability condition on g and the absolute continuity of the Lebesgue integral, λ is absolutely continuous with respect to the Lebesgue measure. If one now revisits the proof of Lemma 1.74, it is easily seen that in this situation the subfamily $(B_{\varrho_i}(x_i))_{i\in I}$ of pairwise disjoint balls, which was used to construct the covering of E_j, satisfies

$$\mathcal{L}^n\left(\bigcup_{i\in I} B_{\varrho_i}(x_i)\right) = \sum_{i\in I} \omega_n \varrho_i^n$$
$$\leq \delta^{n-\alpha}\omega_n \sum_{i\in I} \varrho_i^{\alpha}$$
$$\leq \delta^{n-\alpha}\omega_n j\lambda(\Omega) \to 0 \qquad \text{as } \delta \searrow 0$$

for every $j \in \mathbb{N}$. With the analogous calculations as at the end of the proof of Lemma 1.74 and the absolute continuity of λ with respect to the Lebesgue measure, this allows us to deduce

$$\mathcal{H}^{\alpha}_{5\delta}(E_j) \leq \omega_\alpha 5^\alpha j\lambda(\cup_{i\in I} B_{\varrho_i}(x_i)) \to 0 \quad \text{as } \delta \searrow 0$$

for every $j \in \mathbb{N}$, which in turn proves the claim $\mathcal{H}^{\alpha}(E^{\alpha}) = 0$.

Hausdorff dimension of non-Lebesgue points of Sobolev functions
With the previous lemma at hand we next deduce an estimate for the set of non-Lebesgue-points of (classical and fractional) Sobolev functions.

Proposition 1.76 *Consider* $f \in W^{\theta,p}(\Omega, \mathbb{R}^N)$ *for* $\theta \in (0,1]$ *and* $p \in [1,\infty)$ *with* $\theta p < n$. *Moreover, let*

$$A := \left\{ x_0 \in \Omega \colon \limsup_{\varrho \searrow 0} \fint_{\Omega(x_0,\varrho)} \left| f(x) - (f)_{\Omega(x_0,\varrho)} \right|^p dx > 0 \right\},$$

$$B := \left\{ x_0 \in \Omega \colon \limsup_{\varrho \searrow 0} \left| (f)_{\Omega(x_0,\varrho)} \right| = \infty \right\}.$$

Then we have

$$\dim_{\mathcal{H}}(A) \leq n - \theta p \qquad \text{and} \qquad \dim_{\mathcal{H}}(B) \leq n - \theta p.$$

Proof The proof is essentially based on the arguments of Mingione in [63, Section 4]. We first consider the case $\theta \in (0,1)$ and comment at the end of the proof on the easier case $\theta = 1$. We start by defining a set-function λ via

$$\lambda(O) := \int_O \int_O \frac{|f(x) - f(y)|^p}{|x - y|^{n+\theta p}} \, dx \, dy$$

for every open subset $O \subset \Omega$, and we observe that all assumptions on λ in the previous Lemma 1.74 are fulfilled. In order to estimate the Hausdorff dimension of A, we define a set $S_A \subset \Omega$ via

$$S_A := \left\{ x_0 \in \Omega : \limsup_{\varrho \searrow 0} \varrho^{\theta p - n} \lambda(\Omega(x_0, \varrho)) > 0 \right\}.$$

Now let $\varepsilon > 0$. Lemma 1.74 implies $\dim_{\mathcal{H}}(S_A) \leq n - \theta p$, and consequently we have $\mathcal{H}^{n - \theta p + \varepsilon}(S_A) = 0$. We next observe that if $x_0 \in A$, then $x_0 \in S_A$, due to the Poincaré-type inequality in Lemma 1.66 which gives

$$\varrho^{\theta p - n} \int_{B_\varrho(x_0)} \int_{B_\varrho(x_0)} \frac{|f(x) - f(y)|^p}{|x - y|^{n+\theta p}} \, dx \, dy$$

$$\geq c^{-1}(n, p) \fint_{B_\varrho(x_0)} |f(x) - (f)_{B_\varrho(x_0)}|^p \, dx$$

for all $\varrho < \mathrm{dist}(x_0, \partial\Omega)$. Therefore, we have $A \subset S_A$, and the upper bound for the Hausdorff dimension of A follows. To infer the analogous result for the set B, we fix $\varepsilon_0 \in (0, \varepsilon)$ and define

$$S_B := \left\{ x_0 \in \Omega : \limsup_{\varrho \searrow 0} \varrho^{\theta p - n - \varepsilon_0} \lambda(\Omega(x_0, \varrho)) > 0 \right\}.$$

Again, Lemma 1.74 yields $\dim_{\mathcal{H}}(S_B) \leq n - \theta p + \varepsilon_0$, and in turn we have $\mathcal{H}^{n - \theta p + \varepsilon}(S_B) = 0$. In order to prove the inclusion $B \subset S_B$, we consider $x_0 \in \Omega \setminus S_B$ and a radius $R < \min\{\mathrm{dist}(x_0, \partial\Omega), 1\}$. Then, employing Jensen's inequality and the fractional Poincaré inequality in Lemma 1.66, we estimate

$$\left| (f)_{B_{2^{-j-1}R}(x_0)} - (f)_{B_{2^{-j}R}(x_0)} \right|^p$$

$$\leq 2^n \fint_{B_{2^{-j}R}(x_0)} \left| f - (f)_{B_{2^{-j}R}(x_0)} \right|^p \, dx$$

$$\leq c(n,p)(2^{-j}R)^{\theta p-n} \int_{B_{2^{-j}R}(x_0)} \int_{B_{2^{-j}R}(x_0)} \frac{|f(x)-f(y)|^p}{|x-y|^{n+\theta p}} \, dx \, dy$$

$$= c(n,p)(2^{-j}R)^{\varepsilon_0}(2^{-j}R)^{\theta p-n-\varepsilon_0} \lambda\big(B_{2^{-j}R}(x_0)\big) \leq c(n,p)2^{-j\varepsilon_0}$$

for every $j \in \mathbb{N}_0$ sufficiently large (depending on the choice x_0). Summing these terms up, we finally obtain

$$\lim_{j\to\infty} |(f)_{B_{2^{-j}R}(x_0)}| \leq c(n,p,\varepsilon_0) < \infty,$$

and thus $x_0 \in \Omega \setminus B$. Since $\varepsilon_0 \in (0,\varepsilon)$ was arbitrary, $\mathcal{H}^{n-\theta p+\varepsilon}(B) = 0$ follows, and the proof of the proposition is complete for $\theta \in (0,1)$.

For $\theta = 1$ the assertion follows exactly as above, but in this case we consider the (simpler and even additive) set function λ defined via

$$\lambda(O) := \int_O |Df(x)|^p \, dx$$

for all open subsets $O \subset \Omega$ (cp. Remark 1.75). We may then use the classical Poincaré inequality in Lemma 1.56 for the Sobolev spaces $W^{1,p}(\Omega,\mathbb{R}^N)$ (instead of the Poincaré inequality for the fractional Sobolev spaces) in order to prove the corresponding inclusions $A \subset S_A$ and $B \subset S_B$, and the assertions on the Hausdorff dimensions of the sets S_A and S_B are again available from Lemma 1.74. $\square$

Chapter 2
Introduction to the Setting

In these lecture notes, our main interest concerns elliptic partial differential equations of second order in divergence form

$$- \operatorname{div} a(x, u, Du) = a_0(x, u, Du) \qquad \text{in } \Omega, \tag{2.1}$$

where $a \colon \Omega \times \mathbb{R}^N \times \mathbb{R}^{Nn} \to \mathbb{R}^{Nn}$ and $a_0 \colon \Omega \times \mathbb{R}^N \times \mathbb{R}^{Nn} \to \mathbb{R}^N$ are given vector fields (called the principal part and the inhomogeneity, respectively), $u \colon \Omega \to \mathbb{R}^N$ is a "solution function" (to be specified below), and Ω is a bounded, open set in $\mathbb{R}^n$ (with $n \geq 2$ and $N \geq 1$). Such a (system of) partial differential equation is classified as *quasilinear* since it is linear in the highest (i.e. second) order derivatives of the unknown function u, with coefficients depending only on the independent space variable x and lower order derivatives of u. Supposing suitable assumptions on the vector fields a and a_0 we wish to investigate the regularity of solutions to (2.1). For this purpose, we will distinguish the *scalar case* $N = 1$ of a single equation and solutions with values in $\mathbb{R}$ (see Chap. 3), and the *vectorial case* of $N > 1$ coupled equations and solutions with values in $\mathbb{R}^N$ (see Chaps. 4 and 5). These two cases exhibit fundamental differences with respect to the regularity properties of their solutions.

In this chapter, we first introduce the concept of weak solutions for equations of the form (2.1) and motivate some elementary assumptions (concerning measurability, growth and ellipticity). Then we comment on the connection to the minimization of variational functionals of the form

$$F[w; \Omega] := \int_\Omega f(x, w, Dw) \, dx \tag{2.2}$$

with an integrand $f \colon \Omega \times \mathbb{R}^N \times \mathbb{R}^{Nn} \to \mathbb{R}$, among all functions $w \colon \Omega \to \mathbb{R}^N$ in a given Dirichlet class, via the Euler–Lagrange formalism.

© Springer International Publishing Switzerland 2016
L. Beck, *Elliptic Regularity Theory*, Lecture Notes of the Unione Matematica Italiana 19, DOI 10.1007/978-3-319-27485-0_2

Weak solutions The idea for the concept of a weak solution to (2.1) departs from the well-known concept of classical solutions.

Definition 2.1 A function $u \in C^2(\Omega, \mathbb{R}^N)$ is called a *classical solution* of the system (2.1) if the identity

$$-\text{div } a(x, u(x), Du(x)) = a_0(x, u(x), Du(x))$$

is satisfied for every point $x \in \Omega$.

If the vector field a is smooth and if u is a classical solution of class C^2 of (2.1), then we can multiply the equation (2.1) by a smooth function φ with compact support in Ω and integrate over Ω. Applying the integration by parts formula, we thus find

$$\int_\Omega a(x, u, Du) \cdot D\varphi \, dx = \int_\Omega a_0(x, u, Du) \cdot \varphi \, dx. \qquad (2.3)$$

This identity might be satisfied for functions u that are not of class C^2 and not even differentiable in the classical sense, and – similarly to the concept of weak derivatives – we can take the validity of (2.3) for all test functions $\varphi \in C_0^\infty(\Omega, \mathbb{R}^N)$ in order to give a quite general definition of weak solutions (which are also called generalized solutions or solutions in the sense of distributions) in a suitable Sobolev space.

Definition 2.2 A function $u \in W_{\text{loc}}^{1,p}(\Omega, \mathbb{R}^N) \cap L_{\text{loc}}^q(\Omega, \mathbb{R}^N)$, for some $p \in [1, \infty)$ and $q \in [p, \infty]$, is called a *weak solution* of the system (2.1) if

$$x \mapsto a(x, u(x), Du(x)) \in L_{\text{loc}}^{p'}(\Omega, \mathbb{R}^{Nn}),$$

$$x \mapsto a_0(x, u(x), Du(x)) \in L_{\text{loc}}^{q'}(\Omega, \mathbb{R}^N),$$

and if for all $\varphi \in C_0^\infty(\Omega, \mathbb{R}^N)$ there holds

$$\int_\Omega a(x, u, Du) \cdot D\varphi \, dx = \int_\Omega a_0(x, u, Du) \cdot \varphi \, dx. \qquad (2.4)$$

Remark 2.3 In fact, weaker integrability assumptions are sufficient to give a meaning to the integrals appearing in (2.4). This leads to the concept of *very weak solutions* which a priori belong only to a Sobolev spaces $W^{1,r}(\Omega, \mathbb{R}^N)$ with $r \in [\min\{1, p-1\}, p)$, cf. [55].

The integrability conditions in Definition 2.2 can actually be guaranteed by suitable assumptions on the vector fields a and a_0. To this end, since u and Du are in general merely measurable functions, we need to ensure in the first place that the compositions $x \mapsto a(x, u(x), Du(x))$ and $x \mapsto a_0(x, u(x), Du(x))$ are again measurable in Ω. In this regard, we shall always suppose that

$a\colon \Omega \times \mathbb{R}^N \times \mathbb{R}^{Nn} \to \mathbb{R}^{Nn}$ and $a_0\colon \Omega \times \mathbb{R}^N \times \mathbb{R}^{Nn} \to \mathbb{R}^N$ satisfy the *Carathéodory condition*, that is, they are measurable with respect to x for all (u,z) and continuous with respect to (u,z) for almost every x (with u denoting the function variable and with z denoting the gradient variable). This is a crucial assumption since the composition of a Carathéodory function with a measurable function is again measurable (note that measurability is not necessarily the case for the composition of two functions which are merely Lebesgue-measurable).

Lemma 2.4 *Let $k \in \mathbb{N}$ and consider a Carathéodory function $h\colon \Omega \times \mathbb{R}^k \to \mathbb{R}$, that is, it satisfies*

(i) $x \mapsto h(x,y)$ *is measurable for every $y \in \mathbb{R}^k$,*
(ii) $y \mapsto h(x,y)$ *is continuous for almost every $x \in \Omega$.*

If $v\colon \Omega \to \mathbb{R}^k$ is a measurable function, then also $x \mapsto h(x,v(x))$ is measurable in Ω.

Proof Following [40, Proof of Lemma 4.2] we first assume that v is a simple function, i.e., that it has a representation

$$v(x) = \sum_{i=1}^{m} \lambda_i \mathbb{1}_{E_i}(x)$$

for measurable, disjoint subsets $(E_i)_{i=1,\dots,m}$ of Ω with $\cup_{i=1}^m E_i = \Omega$ and values $(\lambda_i)_{i=1,\dots,m}$ in $\mathbb{R}$, for some $m \in \mathbb{N}$. Then the sub-level sets to level $\ell \in \mathbb{R}$ of the composition $x \mapsto h(x,v(x))$ are given by

$$\{x \in \Omega\colon h(x,v(x)) < \ell\} = \bigcup_{i=1}^{m} \{x \in E_i\colon h(x,\lambda_i) < \ell\},$$

and as the union of measurable sets all sub-level sets are again measurable. Thus, the composition $x \mapsto h(x,v(x))$ is measurable for every simple function v on Ω. Since an arbitrary measurable function v is the pointwise limit of a sequence $(v_j)_{j\in\mathbb{N}}$ of simple functions, the continuity of h with respect to the y-variable yields

$$h(x,v(x)) = \lim_{j\to\infty} h(x,v_j(x)) \qquad \text{for almost all } x \in \Omega.$$

Consequently, we obtain the measurability of the composition $x \mapsto h(x,v(x))$ for arbitrary measurable functions v on Ω, because it can be written as the pointwise limit of a sequence of measurable functions. □

In the second place, we need to ensure the integrability condition of the compositions $x \mapsto a(x,u(x),Du(x))$ and $x \mapsto a_0(x,u(x),Du(x))$ with suitable exponents p' and q', respectively. In the most classical case $p = q$, this can easily be verified by imposing the following *growth assumptions*

$$|a_0(x, u, z)| \le L(1 + |z|)^{p-1} \tag{2.5}$$

$$|a(x, u, z)| \le L(1 + |z|)^{p-1} \tag{2.6}$$

for some $L \ge 1$, almost every $x \in \Omega$, and all $(u, z) \in \mathbb{R}^N \times \mathbb{R}^{Nn}$.

Remark 2.5 If u is a weak solution of the system (2.1) under the assumptions that the vector fields a and a_0 satisfy the Carathéodory condition and the growth assumptions (2.5) and (2.6) and if u belongs even to $W^{1,p}(\Omega, \mathbb{R}^N)$, then the identity (2.4) holds by approximation for all test functions $\varphi \in W_0^{1,p}(\Omega, \mathbb{R}^N)$. In particular, this allows to use modifications of the weak solution as test functions.

In this situation, every classical solution is clearly a weak solution, but the existence of classical solutions is hard to establish (and in fact fails already for quite simple examples). However, the existence of weak solutions can be obtained under quite general assumptions, for example, via methods from functional analysis like Galerkin's method for nonlinear monotone operators. In a second step, since weak solutions are not a priori differentiable in the classical sense, the regularity of weak solutions needs to be investigated. Both for the existence and regularity theory, an *ellipticity assumption* on the principal part a plays a crucial role. The system (2.1) of partial differential equations is called elliptic if $z \mapsto a(x, u, z)$ is differentiable for almost all $x \in \Omega$ and every $u \in \mathbb{R}^N$ and if the bilinear form $D_z a(x, u, z)$ is positive definite for almost every $x \in \Omega$ and all $(u, z) \in \mathbb{R}^N \times \mathbb{R}^{Nn}$. More precisely, we shall work under a strict ellipticity condition, which is compatible with the integrability assumptions, namely that

$$D_z a(x, u, z)\xi \cdot \xi \ge (1 + |z|)^{p-2}|\xi|^2$$

holds for almost every $x \in \Omega$, all $u \in \mathbb{R}^N$ and all $z, \xi \in \mathbb{R}^{Nn}$. The most prominent example, which satisfies all aforementioned assumptions with $p = 2$ is the Laplace equation, and in the course of the lecture notes we will in particular be able study the regularity of weak solutions to nonlinear variants of this very classical equation.

Minimizers A related problem is the minimization of the variational functional (2.2) in Sobolev spaces (and not in classes of functions which are differentiable in the classical sense).

Definition 2.6 A function $u \in W^{1,p}(\Omega, \mathbb{R}^N)$, for some $p \in [1, \infty)$, is called a *minimizer* of the functional (2.2) in $W^{1,p}(\Omega, \mathbb{R}^N)$ if

$$x \mapsto f(x, u(x), Du(x)) \in L^1(\Omega),$$

and if for all $\varphi \in W_0^{1,p}(\Omega, \mathbb{R}^N)$ we have

$$F[u; \Omega] \leq F[u + \varphi; \Omega].$$

Remark 2.7 Under further regularity assumptions on f (such as Lipschitz-continuity in the gradient variable) one can introduce *weak minimizers*, for which only

$$\int_\Omega [f(x, u + \varphi, Du + D\varphi) - f(x, u, Du)] \, dx \geq 0$$

is required, but not finiteness of $F[u; \Omega]$ (in fact, the existence of the previous integral might follow from cancellation effects), cf. [49].

Similarly as for the quasilinear systems above, the integrability condition in Definition 2.6 can be guaranteed by suitable assumptions on the integrand f. For simplicity, we will always suppose that $f: \Omega \times \mathbb{R}^N \times \mathbb{R}^{Nn} \to \mathbb{R}$ satisfies the Carathéodory condition and the growth assumption

$$|f(x, u, z)| \leq L(1 + |z|)^p \tag{2.7}$$

for some $L \geq 1$, almost every $x \in \Omega$, and all $(u, z) \in \mathbb{R}^N \times \mathbb{R}^{Nn}$. In this way, we do not only ensure that the function $x \mapsto f(x, w(x), Dw(x))$ is in $L^1(\Omega)$ for every function w in the Sobolev space $W^{1,p}(\Omega, \mathbb{R}^N)$, but we also have variational methods at our disposal. In particular, if one requires, in analogy to the introduction of the ellipticity condition for quasilinear systems, in addition coercivity of the integrand in the sense of

$$f(x, u, z) \geq |z|^p$$

for almost every $x \in \Omega$ and all $(u, z) \in \mathbb{R}^N \times \mathbb{R}^{Nn}$, and a suitable notion of convexity of f with respect to the gradient variable (such as convexity or quasiconvexity), one easily obtains the existence of minimizers in arbitrary Dirichlet classes in $W^{1,p}(\Omega, \mathbb{R}^N)$, via the application of the direct method in the calculus of variations.

In order to make a connection to quasilinear elliptic systems in divergence form, we observe that, if f is sufficiently regular, then every minimizer of $F[\,\cdot\,; \Omega]$ is a weak solution of the associated Euler–Lagrange equation. This is essentially a consequence of the fact that the function $t \mapsto F[u + t\varphi; \Omega]$ (with $t \in \mathbb{R}$) attains its minimal value at $t = 0$, and for a rigorous proof we refer to [13, Theorem 3.37].

Lemma 2.8 *Let $u \in W^{1,p}(\Omega, \mathbb{R}^N)$, for some $p \in [1, \infty)$, be a minimizer of the functional (2.2) with a Carathéodory integrand $f: \Omega \times \mathbb{R}^N \times \mathbb{R}^{Nn} \to \mathbb{R}$*

satisfying (2.7) and such that $u \mapsto f(x, u, z)$ *and* $z \mapsto f(x, u, z)$ *are differentiable in the classical sense with*

$$|D_u f(x, u, z)| \leq L(1 + |z|)^{p-1},$$
$$|D_z f(x, u, z)| \leq L(1 + |z|)^{p-1},$$

for almost every $x \in \Omega$, *and all* $(u, z) \in \mathbb{R}^N \times \mathbb{R}^{Nn}$. *Then u is a weak solution of the* Euler–Lagrange system

$$\operatorname{div} D_z f(x, u, Du) = D_u f(x, u, Du) \qquad in\ \Omega.$$

However, often it is not possible to obtain the relevant properties of minimizers via the Euler–Lagrange system. On the one hand, the integrand might be not sufficiently regular for the Euler–Lagrange system to exist. On the other hand, in the passage from the minimization problem to the Euler–Lagrange system one loses the crucial information of the minimization property. As a matter of fact, minimizers enjoy often better regularity properties than a general critical point of the functional, that is, of a general weak solution of the Euler–Lagrange system. For these reasons, parts of regularity theory are treated separately for minimizers of variational functionals and for weak solutions to partial differential equations in divergence form, but nevertheless they share some fundamental features, which will be detailed for two specific examples.

Chapter 3
The Scalar Case

The aim of this chapter is to discuss some full (that is everywhere) regularity results for scalar-valued weak solutions to second order elliptic equations in divergence forms. This means that we study equations of the form

$$- \operatorname{div} a(x, u, Du) = a_0(x, u, Du) \qquad \text{in } \Omega, \qquad (3.1)$$

with $\Omega \subset \mathbb{R}^n$ a bounded domain and where we shall always suppose that the vector field $a \colon \Omega \times \mathbb{R} \times \mathbb{R}^n \to \mathbb{R}^n$ and the inhomogeneity $a_0 \colon \Omega \times \mathbb{R} \times \mathbb{R}^n \to \mathbb{R}$ satisfy the Carathéodory condition and that a suitable ellipticity condition holds. Weak solutions only belong to some Sobolev space (consequently, neither are they necessarily continuous nor do derivatives a priori exist in the classical sense), and hence, their regularity needs to be investigated. In what follows, we prove local Hölder regularity of weak solution to (3.1), merely under these quite general assumptions. For this purpose, we explain two different (and classical) strategies of proof dating back to the late 1950s. First, we present De Giorgi's level set technique developed in [15], in a unified approach that applies both to weak solutions of elliptic equations and to minimizers of variational integrals, via the study of Q-minimizers of suitable functionals. We then address, for the specific case of linear elliptic equations, an alternative proof of the everywhere regularity result of weak solutions via Moser's iteration method relying on [67].

3.1 De Giorgi's Level Set Technique

De Giorgi proved in [15] local Hölder regularity for weak solutions to linear equations of the form

$$\operatorname{div}(a(x)Du) = 0 \qquad \text{in } \Omega,$$

© Springer International Publishing Switzerland 2016
L. Beck, *Elliptic Regularity Theory*, Lecture Notes of the Unione
Matematica Italiana 19, DOI 10.1007/978-3-319-27485-0_3

assuming merely that the coefficients $a\colon \Omega \to \mathbb{R}^{n\times n}$ are measurable, bounded and elliptic. With this initial regularity result at hand, one easily finds that weak solutions of such equations are in fact analytic whenever the equation has analytic coefficients. In doing so De Giorgi gave an affirmative answer to one of the celebrated 23 problems, which were outlined by Hilbert [47] at the International Congress of Mathematicians in Paris in 1900, namely to

Hilbert's 19th Problem: "Sind die Lösungen regulärer Variationsprobleme stets notwendig analytisch?" (Are the solutions of regular variational problems always analytic?)

To provide some further historical background, let us note that Hilbert was particularly interested in the minimization of convex variational functionals (and here in the most relevant case $n = 2$ and $N = 1$ from the point of view of physics). Restricting ourselves for the moment to simple functionals of the form

$$w \mapsto \int_\Omega f(Dw)\, dx$$

for a smooth, convex integrand $f\colon \mathbb{R}^n \to \mathbb{R}$, the associated Euler–Lagrange equation, cf. Lemma 2.8, is a homogeneous elliptic partial differential equation with a smooth vector field. Moreover, formal differentiation shows that every minimizer u solves the equation

$$\operatorname{div}\left(D_z^2 f(Du) D D_i u\right) = 0 \qquad \text{in } \Omega$$

for every $i \in \{1, \ldots, n\}$. In other words, every partial first order derivative of u solves a linear elliptic equation, with coefficients given by $a(x) := D_z^2 f(Du(x))$ for $x \in \Omega$. Therefore, if one starts from a minimizer, which a priori belongs only to some Sobolev space, then the coefficients of the (linear) equation solved by its derivatives are merely measurable, which is precisely the assumption imposed originally by De Giorgi. We further remark that Nash [69] obtained independently and simultaneously to De Giorgi a similar result, which states the regularity of bounded weak solutions to parabolic equations (and, by specializing in the time-independent case, also to elliptic equations). Shortly after the publications of De Giorgi and Nash, Moser [67] proposed another strategy of proof to deduce everywhere regularity of weak solutions, which will be detailed in Sect. 3.2. Later on, these results were extended to weak solutions of nonlinear equations and to minimizers of variational functionals, still under very mild assumptions (see for example the book of Ladyzhenskaya and Ural'tseva [53] or the paper [32] by Giaquinta and Giusti). In conclusion, Hilbert's 19th problem is considered to be solved.

In this section we present the essential ideas of De Giorgi's method, in a unified approach that applies both to weak solutions of elliptic equations and to minimizers of variational integrals, via the study of Q-minimizers of suitable functionals. We follow closely the exposition in Giusti's monograph [40, Chapter 7] (where even more general functionals are discussed). Before explaining the idea behind, we start by introducing the concept and some examples of Q-minimizers.

Definition 3.1 Let $Q \geq 1$. A function $u \in W^{1,p}(\Omega)$, for some $p \in [1, \infty)$, is called a *Q-minimizer* of the functional (2.2) in $W^{1,p}(\Omega)$ if $x \mapsto f(x, u(x), Du(x)) \in L^1(\Omega)$ and if for all open sets $\Omega' \subset \Omega$ and every $\varphi \in W_0^{1,p}(\Omega')$ we have

$$F[u; \Omega'] \leq QF[u + \varphi; \Omega'].$$

If the previous inequality holds only for all non-positive or all non-negative functions $\varphi \in W_0^{1,p}(\Omega')$, then u is called a *sub-Q-minimizer* or *super-Q-minimizer*, respectively.

Remarks 3.2

(i) The notion of Q-minimizers was introduced by Giaquinta and Giusti in [34] in order to unify the treatment of weak solutions to elliptic equations in divergence form and of minimizers of variational integrals (and other related problems).

(ii) Minimizers are obviously Q-minimizers for $Q = 1$ (but not necessarily for any $Q > 1$, as can be easily demonstrated by taking the integrand $f \equiv -1$).

(iii) Another definition of Q-minimizer requires the minimality condition only for $\Omega' = \operatorname{spt} \varphi$. At least in the case $f \geq 0$, this definition is easily seen to be equivalent. In our later consideration, this will always be the case (otherwise $F[u + \varphi; \Omega']$ might be negative and then the existence of Q-minimizers with $Q > 1$ could fail). Moreover, the assumption of non-negativity of the integrand implies that every Q-minimizer is in particular a Q'-minimizer for every $Q' \geq Q$.

Examples 3.3 (of Q-minimizers)

(i) *Assume that the integrand $f: \Omega \times \mathbb{R} \times \mathbb{R}^n \to \mathbb{R}$ satisfies the Carathéodory condition and suppose further that the coercivity and growth condition $|z|^p \leq f(x, u, z) \leq L|z|^p$ holds for almost every $x \in \Omega$, all $(u, z) \in \mathbb{R} \times \mathbb{R}^n$ and some $p \in [1, \infty)$. Then every minimizer $u \in W^{1,p}(\Omega)$ of $F[\,\cdot\,; \Omega]$ is a Q-minimizer (with $Q = L$) of the* Dirichlet *or p-energy*

$$E_p[w; \Omega] := \int_\Omega |Dw|^p \, dx.$$

Accordingly, every Q_f-minimizer of $F[\,\cdot\,; \Omega]$ is a Q-minimizer (with $Q = LQ_f$) of $E_p[\,\cdot\,; \Omega]$.

(ii) *Assume that the vector field* $a\colon \Omega \times \mathbb{R} \times \mathbb{R}^n \to \mathbb{R}^n$ *and the inhomogeneity*
$a_0\colon \Omega \times \mathbb{R} \times \mathbb{R}^n \to \mathbb{R}$ *satisfy the Carathéodory condition, the growth*
conditions (2.5) and (2.6), and suppose further that $a(x,u,z) \cdot z \geq |z|^p$
holds for almost every $x \in \Omega$, *all* $(u,z) \in \mathbb{R} \times \mathbb{R}^n$ *and some* $p \in (1,\infty)$.
Then every weak solution $u \in W^{1,p}(\Omega)$ *to the equation (3.1) is a*
Q-minimizers of the functional

$$\tilde{E}_p[w;\Omega] := \int_\Omega (1 + |Dw|)^p \, dx$$

with $Q = Q(n,p,L,\Omega, \|Du\|_{L^p(\Omega,\mathbb{R}^n)})$.

Proof We consider an arbitrary function $\varphi \in W_0^{1,p}(\Omega')$, for Ω' an open subset
of Ω.

(i) Since $F[u+\varphi;\Omega\setminus\Omega'] = F[u;\Omega\setminus\Omega']$ holds, the assertion follows from the
Q_f-minimizing property of u, combined with the growth assumptions on
the integrand:

$$\int_{\Omega'} |Du|^p \, dx \leq \int_{\Omega'} f(x,u,Du)\, dx$$

$$\leq Q_f \int_{\Omega'} f(x,u+\varphi,Du+D\varphi)\,dx$$

$$\leq LQ_f \int_{\Omega'} |Du+D\varphi|^p\, dx\,.$$

(ii) Since u is a weak solution, we first observe

$$\int_{\Omega'} a(x,u,Du)\cdot Du\, dx$$

$$= \int_{\Omega'} a(x,u,Du)\cdot(Du+D\varphi)\,dx - \int_{\Omega'} a_0(x,u,Du)\varphi\, dx\,.$$

Hence, invoking the growth assumptions on a_0, a and applying Hölder's
inequality, we find

$$\int_{\Omega'} |Du|^p\, dx$$

$$\leq L\int_{\Omega'} (1+|Du|)^{p-1}|Du+D\varphi|\,dx + L\int_{\Omega'}(1+|Du|)^{p-1}|\varphi|\,dx$$

$$\leq c(p,L)\Big(\int_{\Omega'}(1+|Du|)^p\,dx\Big)^{\frac{p-1}{p}}\Big(\int_{\Omega'}(|Du+D\varphi|^p+|\varphi|^p)\,dx\Big)^{\frac{1}{p}}.$$

To estimate the integral involving φ on the right-hand side, we use Poincaré's inequality from Lemma 1.56 and obtain

$$\int_{\Omega'} |\varphi|^p \, dx \leq c(n, p, \Omega) \int_{\Omega'} \left(|Du|^p + |Du + D\varphi|^p \right) dx.$$

Combining these two inequalities, we arrive at the assertion.   □

De Giorgi's method is based on the geometric idea that boundedness (and in a second step even continuity) of a measurable function u, defined over Ω, can be investigated via the analysis of the decay of its level sets. For a ball $B_R(x_0) \subset \Omega$, we hence introduce the *super-level set* and the *sub-level set* of u to level $k \in \mathbb{R}$ as

$$A(k, x_0, R) := \left\{ x \in B_R(x_0) \colon u(x) > k \right\},$$
$$B(k, x_0, R) := \left\{ x \in B_R(x_0) \colon u(x) < k \right\}.$$

Obviously, we have $|A(k, x_0, R)| + |B(k, x_0, R)| = |B_R(x_0)|$ for almost every level k, and the super-level set of u to the level k is exactly the sub-level set of the function $-u$ to the level $-k$. This simple fact relates sub-Q-minimizers of F to super-Q-minimizers of the related functional with integrand given by $f(x, -u, -z)$ and vice versa. In what follows, we will precisely study the decay of the level set in dependence of the level k, and we start by explaining, how sub-Q-minimality and super-Q-minimality allows for estimates of the measure of $A(k, x_0, R)$ and $B(k, x_0, R)$, respectively.

3.1.1 Local Boundedness

Our first aim is to show that every sub-Q-minimizer is bounded from above and that every super-Q-minimizer is bounded from below, supposing only p-growth assumptions on the integrand f of the functional F defined in (2.2). To this end, we observe that one-sided boundedness is equivalent to the fact that the super-level set $A(k, x_0, R)$ and the sub-level set $B(-k, x_0, R)$, respectively, are of Lebesgue measure zero for some finite number k. If the function under consideration belongs to the Sobolev space $W^{1,p}(\Omega)$ for some $p > n$, then the existence of such a level k is of course trivial, due to Morrey's embedding from Theorem 1.61, but for the general case $p \geq 1$, this is a non-trivial task, which is accomplished in this section (along with a quantitative estimate). The starting point for the analysis of the level sets is the following lemma.

Lemma 3.4 *Let $u \in W^{1,p}(\Omega)$, for some $p \in [1, \infty)$, be a sub-Q-minimizer of the functional $F[\,\cdot\,; \Omega]$ with a Carathéodory integrand $f \colon \Omega \times \mathbb{R} \times \mathbb{R}^n \to \mathbb{R}$*

satisfying the growth condition

$$|z|^p \le f(x, u, z) \le L(1 + |z|)^p \tag{3.2}$$

for almost every $x \in \Omega$ and all $(u, z) \in \mathbb{R} \times \mathbb{R}^n$. There exists a constant c depending only on p, L, and Q such that, for every $k \in \mathbb{R}$ and every pair of concentric balls $B_r(x_0) \Subset B_R(x_0) \subset \Omega$, we have

$$\int_{A(k,x_0,r)} |Du|^p \, dx \le c(R-r)^{-p} \int_{A(k,x_0,R)} (u-k)^p \, dx + c|A(k,x_0,R)| \, .$$

Proof Without loss of generality we assume $x_0 = 0$. We consider $r \le \varrho < \sigma \le R$ and take a cut-off function $\eta \in C_0^\infty(B_\sigma, [0,1])$ satisfying $\eta \equiv 1$ in B_ϱ and $|D\eta| \le 2(\sigma - \varrho)^{-1}$. We now use the sub-$Q$-minimizing property of u with the test function $\varphi = -\eta(u - k)_+$, which means that we compare u with a modification of u obtained by essentially cutting off the values greater than k in the smaller ball B_ϱ (and interpolating between u and the cut version in the annulus $B_\sigma \setminus B_\varrho$). By definition of the set $A(k, 0, \sigma)$ it is clear that φ is different from zero at most on $A(k, 0, \sigma)$. Note that a priori this set is only measurable, but not necessarily open. However, due to the growth assumptions on f and the absolute continuity of the integral, the domain of integration in the definition of (sub- or super-) Q-minimizers may also be chosen as a measurable set containing the support of φ. In this way, we find

$$\int_{A(k,0,\sigma)} |Du|^p \, dx \le \int_{A(k,0,\sigma)} f(x, u, Du) \, dx$$

$$\le Q \int_{A(k,0,\sigma)} f(x, u + \varphi, D(u + \varphi)) \, dx$$

$$\le LQ \int_{A(k,0,\sigma)} (1 + |D(u + \varphi)|)^p \, dx \, .$$

To estimate the right-hand side we observe

$$|D(u + \varphi)|^p \le c(p)\big((1 - \eta)^p |Du|^p + (\sigma - \varrho)^{-p}(u - k)^p\big)$$

on $A(k, 0, \sigma)$. With $\eta \equiv 1$ on B_ϱ, this gives

$$\int_{A(k,0,\varrho)} |Du|^p \, dx \le c_0(p, L, Q) \bigg[\int_{A(k,0,\sigma)} \big(1 + (\sigma - \varrho)^{-p}(u - k)^p\big) \, dx$$

$$+ \int_{A(k,0,\sigma) \setminus A(k,0,\varrho)} |Du|^p \, dx \bigg] \, .$$

Now we *fill the hole* in the second integral on the right-hand side by adding the integral of $c_0|Du|^p$ over $A(k, 0, \varrho)$ to both sides of the inequality (and with c_0 exactly the constant appearing in the previous inequality). This yields after division by $c_0 + 1$:

$$\int_{A(k,0,\varrho)} |Du|^p \, dx \le \int_{A(k,0,\sigma)} \left(1 + (\sigma - \varrho)^{-p}(u-k)^p \right) dx$$
$$+ \frac{c_0}{c_0+1} \int_{A(k,0,\sigma)} |Du|^p \, dx \, .$$

Recalling that the radii ϱ and σ were chosen arbitrarily with $r \le \varrho < \sigma \le R$, we may employ the iteration Lemma B.1 with the choices

$$\phi(\varrho) := \int_{A(k,0,\varrho)} |Du|^p \, dx \quad \text{and} \quad \vartheta = c_0/(c_0+1) \in (0,1) \, .$$

This yields

$$\int_{A(k,0,r)} |Du|^p \, dx \le c(p, L, Q)(R-r)^{-p} \int_{A(k,0,R)} (u-k)^p \, dx + c|A(k,0,R)|$$

and finishes the proof of the lemma.   □

Corollary 3.5 *Let $u \in W^{1,p}(\Omega)$, for some $p \in [1,\infty)$, be a super-Q-minimizer of the functional $F[\,\cdot\,; \Omega]$ with a Carathéodory integrand $f: \Omega \times \mathbb{R} \times \mathbb{R}^n \to \mathbb{R}$ satisfying the growth condition (3.2). There exists a constant c depending only on p, L, and Q such that, for every $k \in \mathbb{R}$ and every pair of concentric balls $B_r(x_0) \Subset B_R(x_0) \subset \Omega$, we have*

$$\int_{B(k,x_0,r)} |Du|^p \, dx \le c(R-r)^{-p} \int_{B(k,x_0,R)} (k-u)^p \, dx + c|B(k,x_0,R)| \, .$$

Proof If u is a super-Q-minimizer of the functional $F[\,\cdot\,; \Omega]$, then $-u$ is a sub-Q-minimizer of the functional $F'[w; \Omega] := \int_\Omega f(x, -w, -Dw) \, dx$. With the substitutions u by $-u$ and k by $-k$ the assertion follows from Lemma 3.4. □

Remark 3.6 The *hole-filling technique* was first implemented by Widman [83]. It is nowadays a standard tool in the regularity theory for parabolic and elliptic problems, which allows to obtain immediately – and without the application of further deep results such as Gehring's Theorem 1.22 – improved estimates and higher regularity of solutions (such as Morrey or Hölder regularity and higher integrability).

Remark 3.7 Estimates of the form

$$\|Du\|_{L^p(O',\mathbb{R}^n)} \leq c(O',O)\|u\|_{L^p(O)} + \text{error terms}$$

for sets $O' \subset O$ are called *Caccioppoli-type inequalities*, named after Caccioppoli who established similar inequalities for weak solutions of elliptic boundary value problems in [7]. These inequalities are some sort of reverse Poincaré inequality, and for what concerns regularity theory they are usually established as a first step. In this sense the estimates stated in Lemma 3.4 and Corollary 3.5 represent Caccioppoli-type inequalities on level sets.

De Giorgi's achievement in [15] was the discovery that all information concerning boundedness (and even Hölder continuity as we shall see later) from below or from above is encoded in the previous Caccioppoli-type inequalities. Therefore, he introduced new classes of functions – today known as De Giorgi classes –, which are defined via the validity of these inequalities, cp. [40, Definition 7.1] and [53, Chapter 2.5]. However, as a matter of fact, these classes may in general contain also other functions than sub-Q-minimizers or super-Q-minimizers.

Definition 3.8 We say that a function $u \in W^{1,p}(\Omega)$, with $p \in [1,\infty)$, belongs to the *De Giorgi class* $DG_p^+(\Omega)$ if there exists a constant C_0, a number $k_0 \in \mathbb{R}$, and a radius R_0 such that for every pair of concentric balls $B_r(x_0) \Subset B_R(x_0) \subset \Omega$ with $R \leq R_0$ and for every level $k \geq k_0$ we have

$$\int_{A(k,x_0,r)} |Du|^p \, dx \leq C_0(R-r)^{-p} \int_{A(k,x_0,R)} (u-k)^p \, dx + C_0 |A(k,x_0,R)|.$$

We further say that u belongs to $DG_p^-(\Omega)$ if $-u$ belongs to the De Giorgi class $DG_p^+(\Omega)$. Finally, we define $DG_p(\Omega)$ as the class of functions in $W^{1,p}(\Omega)$ which belong to both De Giorgi classes $DG_p^+(\Omega)$ and $DG_p^-(\Omega)$.

Via Sobolev's inequality as the second main ingredient, these Caccioppoli-type inequalities now lead to (one-sided) boundedness of functions in the De Giorgi classes, cf. [15, Lemma IV].

Theorem 3.9 (De Giorgi) *Let $u \in DG_p^+(\Omega)$ for some $p \in [1,\infty)$. Then u is locally bounded from above, i.e. $u_+ \in L_{loc}^\infty(\Omega)$, and for every ball $B_R(x_0) \subset \Omega$ with $R \leq R_0$ we have*

$$\sup_{B_{R/2}(x_0)} u \leq k_0 + c_1(n,p,C_0)R$$

$$+ c_1(n,p,C_0)\left(R^{-n}\int_{A(k_0,x_0,R)} (u-k_0)^p \, dx\right)^{\frac{1}{p}} \left(\frac{|A(k_0,x_0,R)|}{R^n}\right).$$

Proof Without loss of generality we may assume $x_0 = 0$. We fix $R \leq R_0$ such that $B_R \subset \Omega$ and a number $k > k_0$. Then we take two radii ϱ, σ such that $R/2 \leq \varrho < \sigma \leq R$ and a cut-off function $\eta \in C_0^\infty(B_{(\varrho+\sigma)/2}, [0,1])$ satisfying $\eta \equiv 1$ on B_ϱ and $|D\eta| \leq 4(\sigma - \varrho)^{-1}$. Applying first Hölder's inequality, then Sobolev's inequality (Lemma 1.51), and finally the Caccioppoli-type estimate from Definition 3.8 of the De Giorgi class $DG_p^+(\Omega)$, we find for $p < n$

$$\int_{A(k,0,\varrho)} (u-k)^p \, dx \leq \int_{A(k,0,(\varrho+\sigma)/2)} \eta^p (u-k)^p \, dx$$

$$\leq |A(k,0,\sigma)|^{1-\frac{p}{p^*}} \left(\int_{A(k,0,(\varrho+\sigma)/2)} \eta^{p^*} (u-k)^{p^*} \, dx \right)^{\frac{p}{p^*}}$$

$$\leq c(n,p) |A(k,0,\sigma)|^{1-\frac{p}{p^*}} \left[\int_{A(k,0,(\varrho+\sigma)/2)} |Du|^p \, dx \right.$$

$$\left. + (\sigma - \varrho)^{-p} \int_{A(k,0,(\varrho+\sigma)/2)} (u-k)^p \, dx \right]$$

$$\leq c(n,p,C_0) |A(k,0,\sigma)|^{\frac{p}{n}} \left[(\sigma - \varrho)^{-p} \int_{A(k,0,\sigma)} (u-k)^p \, dx + |A(k,0,\sigma)| \right].$$

$$(3.3)$$

This inequality is obtained in a similar way for $p \geq n$, by first applying Sobolev's inequality (with exponent $np/(n+p) \in [1,n)$ instead of p) and then Hölder's inequality. The significance of this inequality becomes clear by looking carefully at the crucial terms arising on the right-hand side of (3.3). We first observe that the same integral as on the left-hand side appears, but on the super-level set for the larger ball B_σ. Secondly, the factor $(\sigma - \varrho)^{-p}$ arises, which is critical only for small values of $\sigma - \varrho$. Finally (and most importantly), the factor $|A(k,0,\sigma)|^{\frac{p}{n}}$ is present, which can be made arbitrarily small for k large. To have a quantitative estimate for this smallness, we calculate, for a level $h \in [k_0, k)$,

$$|A(k,0,\sigma)| \leq (k-h)^{-p} \int_{A(k,0,\sigma)} (u-h)^p \, dx$$

$$\leq (k-h)^{-p} \int_{A(h,0,\sigma)} (u-h)^p \, dx.$$

Hence, the measure of $A(k,0,\sigma)$ is related to $\int_{A(h,0,\sigma)} (u-h)^p \, dx$ via negative powers of the difference $k - h$. Moreover, we easily see

$$\int_{A(k,0,\sigma)} (u-k)^p \, dx \leq \int_{A(k,0,\sigma)} (u-h)^p \, dx \leq \int_{A(h,0,\sigma)} (u-h)^p \, dx.$$

For a parameter $q \geq 0$ we next define a function $\phi \colon [k_0, \infty) \times [R/2, R] \to \mathbb{R}_0^+$ via

$$\phi(k, \varrho) := |A(k, 0, \varrho)|^q \int_{A(k,0,\varrho)} (u - k)^p \, dx$$

(and note that ϕ is non-increasing in k for fixed ϱ and non-decreasing in ϱ for fixed k). Combining the previous inequalities with the estimate (3.3) (multiplied by $|A(k, 0, \varrho)|^q$) we thus obtain

$\phi(k, \varrho)$

$$\leq c(n, p, C_0) |A(k, 0, \sigma)|^{q + \frac{p}{n}} \left[(\sigma - \varrho)^{-p} \int_{A(k,0,\sigma)} (u - k)^p \, dx + |A(k, 0, \sigma)| \right]$$

$$\leq c(n, p, C_0) |A(k, 0, \sigma)|^{q + \frac{p}{n}} \left[(\sigma - \varrho)^{-p} + (k - h)^{-p} \right] \int_{A(h,0,\sigma)} (u - h)^p \, dx$$

$$\leq c(n, p, C_0) \left[(\sigma - \varrho)^{-p} (k - h)^{-\frac{p^2}{n(1+q)}} + (k - h)^{-p - \frac{p^2}{n(1+q)}} \right]$$

$$\times |A(h, 0, \sigma)|^{q + \frac{p}{n} - \frac{p}{n(1+q)}} \left(\int_{A(h,0,\sigma)} (u - h)^p \, dx \right)^{1 + \frac{p}{n(1+q)}}$$

$$= c(n, p, C_0) \left[(\sigma - \varrho)^{-p} (k - h)^{-\frac{p^2}{n(1+q)}} + (k - h)^{-p - \frac{p^2}{n(1+q)}} \right] \phi(h, \sigma)^{1 + \frac{p}{n(1+q)}} .$$

We are now in the position to apply Lemma B.2, with exponents $\alpha_1 = p^2/(n(1+q))$, $\alpha_2 = p$, and $\beta = 1 + p/(n(1+q))$, which implies that $\phi(k, R/2)$ vanishes for $k \geq k_0 + d$ sufficiently great (with d given by Lemma B.2). Equivalently this can be written as

$$\sup_{B_{R/2}} u \leq k_0 + c(n, p, q, C_0) R$$

$$+ c(n, p, q, C_0) R^{-\frac{n(1+q)}{p}} |A(k_0, 0, R)|^{\frac{q}{p}} \left(\int_{A(k_0,0,R)} (u - k_0)^p \, dx \right)^{\frac{1}{p}} ,$$

which (with the choice $q = p$) proves the assertion. $\square$

As an immediate consequence, we obtain for each function $u \in DG_p^-(\Omega)$ (which by definition is equivalent to $-u \in DG_p^+(\Omega)$) the corresponding statement of local boundedness from below.

Corollary 3.10 *Let $u \in DG_p^-(\Omega)$ for some $p \in [1, \infty)$. Then u is locally bounded from below, i.e. $u_- \in L^\infty_{\text{loc}}(\Omega)$, and for every ball $B_R(x_0) \subset \Omega$ with $R \leq R_0$ we have*

$$\inf_{B_{R/2}(x_0)} u \geq -k_0 - c_1(n, p, C_0)R$$

$$- c_1(n, p, C_0)\left(R^{-n}\int_{B(-k_0, x_0, R)}(-k_0 - u)^p\, dx\right)^{\frac{1}{p}}\left(\frac{|B(-k_0, x_0, R)|}{R^n}\right).$$

3.1.2   Local Hölder Continuity

We next study functions belonging to a (suitable) De Giorgi class $DG_p(\Omega)$ and wish to show local Hölder continuity, relying only the quantitative L^∞-estimates derived in the previous section combined with the validity of the Caccioppoli-type inequalities (note that the L^∞-estimate a priori does not exclude discontinuities). To this end, we introduce, for a locally bounded function $u \in L^\infty_{\text{loc}}(\Omega)$, the notations

$$M(x_0, R) := \sup_{B_R(x_0)} u$$

$$m(x_0, R) := \inf_{B_R(x_0)} u$$

$$\text{osc}(x_0, R) := M(x_0, R) - m(x_0, R)$$

for an arbitrary ball $B_R(x_0) \Subset \Omega$. In a first step we investigate the behavior of the size of the super-level sets at levels in a neighbourhood of the maximum of u and show some sort of continuity or quantified smallness of the super-level sets, cf. [40, Lemma 7.2].

Lemma 3.11 *Let $u \in DG_p^+(\Omega) \cap L^\infty_{\text{loc}}(\Omega)$ for some $p \in (1, \infty)$ and let $B_{2R}(x_0) \Subset \Omega$ with $2R \leq R_0$. Assume further that*

$$|A(\bar{k}, x_0, R)| < \gamma|B_R(x_0)|$$

holds for some $\gamma \in (0, 1)$ and $\bar{k} := (M(x_0, 2R) + m(x_0, 2R))/2 \geq k_0$. If for some integer $\ell \in \mathbb{N}$ we have

$$\text{osc}(x_0, 2R) \geq 2^\ell R, \tag{3.4}$$

then there holds

$$|A(k, x_0, R)| \leq c_2(n, p, C_0, \gamma)\ell^{-\frac{n(p-1)}{(n-1)p}}R^n$$

for all levels $k \geq M(x_0, 2R) - 2^{-\ell-1}\text{osc}(x_0, 2R)$.

Proof We here follow [40, Proof of Lemma 7.2]. Without loss of generality we assume $x_0 = 0$. For $M(0, 2R) \geq k > h \geq \bar{k}$ we define a non-negative function v via

$$v := \begin{cases} k - h & \text{if } u \geq k, \\ u - h & \text{if } h < u < k, \\ 0 & \text{if } u \leq h. \end{cases}$$

Since v is identically zero on $(B_R \setminus A(h, 0, R)) \supset (B_R \setminus A(\bar{k}, 0, R))$, it vanishes on a set of measure greater than $(1 - \gamma)|B_R|$. With the same reasoning as in Remark 1.57 (ii) we may thus apply the Sobolev–Poincaré inequality and then Hölder's inequality to obtain

$$(k - h)|A(k, 0, R)|^{1 - \frac{1}{n}} \leq \left(\int_{B_R} v^{\frac{n}{n-1}} \, dx \right)^{1 - \frac{1}{n}}$$

$$\leq c(n, p, \gamma) \int_{B_R} |Dv| \, dx$$

$$= c(n, p, \gamma) \int_{A(h,0,R) \setminus A(k,0,R)} |Du| \, dx$$

$$\leq c(n, p, \gamma)|A(h, 0, R) \setminus A(k, 0, R)|^{1 - \frac{1}{p}} \left(\int_{A(h,0,R)} |Du|^p \, dx \right)^{\frac{1}{p}}.$$

In view of $u \in DG_p^+(\Omega)$ and $h \geq k_0$, the integral on the right-hand side is estimated by

$$\int_{A(h,0,R)} |Du|^p \, dx \leq C_0 R^{-p} \int_{A(h,0,2R)} |u - h|^p \, dx + C_0|A(h, 0, 2R)|$$

$$\leq c(n, C_0) R^{n-p}(M(0, 2R) - h)^p + c(n, C_0) R^n.$$

In combination with the previous inequality, we hence find

$$(k - h)^{\frac{p}{p-1}} |A(k, 0, R)|^{\frac{(n-1)p}{n(p-1)}}$$

$$\leq c|A(h, 0, R) \setminus A(k, 0, R)| \left[R^{\frac{n-p}{p-1}} (M(0, 2R) - h)^{\frac{p}{p-1}} + R^{\frac{n}{p-1}} \right]$$

$$= c \left[|A(h, 0, R)| - |A(k, 0, R)| \right] R^{\frac{n-p}{p-1}} \left[(M(0, 2R) - h)^{\frac{p}{p-1}} + R^{\frac{p}{p-1}} \right]$$

with a constant c depending only on n, p, C_0, and γ. For $i \in \mathbb{N}_0$ we next define an increasing sequence of levels

$$k_i = M(0, 2R) - 2^{-i-1} \operatorname{osc}(0, 2R)$$

and compute

$$M(0, 2R) - k_{i-1} = 2^{-i}\operatorname{osc}(0, 2R) = 2(k_i - k_{i-1}).$$

Applying the previous inequality for the levels $k = k_i$ and $h = k_{i-1}$, we obtain

$$|A(k_\ell, 0, R)|^{\frac{(n-1)p}{n(p-1)}} \leq |A(k_i, 0, R)|^{\frac{(n-1)p}{n(p-1)}}$$

$$\leq c \big[|A(k_{i-1}, 0, R)| - |A(k_i, 0, R)| \big] R^{\frac{n-p}{p-1}}$$

$$\times \big[1 + (2^{i+1}\operatorname{osc}(0, 2R)^{-1}R)^{\frac{p}{p-1}} \big]$$

for all $i \in \{1, \dots, \ell\}$. Summing i from 1 to ℓ and employing assumption (3.4), we get

$$\ell |A(k_\ell, 0, R)|^{\frac{(n-1)p}{n(p-1)}} \leq c|A(k_0, 0, R)|R^{\frac{n-p}{p-1}} \leq c(n, p, C_0, \gamma)R^{\frac{(n-1)p}{p-1}},$$

and the claim follows by monotonicity of the level sets.   □

Taking into account the boundedness result from the previous section, we now proceed to the local Hölder regularity result, cf. [15, Theorema I] and [40, Theorem 7.6].

Theorem 3.12 (De Giorgi) *Let $u \in DG_p(\Omega)$, for some $p \in (1, \infty)$, be a function which satisfies the Caccioppoli-type inequalities in Definition 3.8 for all levels $k \in \mathbb{R}$ and $R_0 = 1$. Then there exists a positive exponent $\alpha = \alpha(n, p, C_0)$ such that u is locally Hölder continuous in Ω with Hölder exponent α, i.e. $u \in C^{0,\alpha}(\Omega)$.*

Proof In view of Theorem 3.9 and Corollary 3.10 u is bounded in every compactly supported subset $\Omega' \Subset \Omega$ with

$$\|u\|_{L^\infty(\Omega')} \leq c(n, p, C_0, \Omega, \Omega')\big(1 + \|u\|_{L^p(\Omega)}\big).$$

Hence, it only remains to bound the $C^{0,\alpha}$-Hölder semi-norm in Ω' for a suitable exponent α (and it is crucial that α does not depend on Ω'). Without loss of generality we take $x_0 = 0$ and consider a ball $B_{2R} \Subset \Omega$ with $2R \leq R_0$. Setting $\bar{k} := (M(0, 2R) + m(0, 2R))/2$ as above, we may further suppose that

$$|A(\bar{k}, 0, R)| \leq \frac{1}{2}|B_R|$$

holds (otherwise we replace u by $-u$). Next we consider the levels $k_i = M(0, 2R) - 2^{-i-1}\operatorname{osc}(0, 2R)$ for $i \in \mathbb{N}_0$. According to the quantitative L^∞-estimate from Theorem 3.9 on $B_{R/2}$, applied for the level k_i instead of k_0, we have

$$\sup_{B_{R/2}} u \leq k_i + c_1(n,p,C_0)R$$

$$+ c_1(n,p,C_0)\left(R^{-n}\int_{A(k_i,0,R)}|u-k_i|^p\,dx\right)^{\frac{1}{p}}\left(\frac{|A(k_i,0,R)|}{R^n}\right)$$

$$\leq k_i + c_1(n,p,C_0)R + \tilde{c}_1(n,p,C_0)\sup_{B_{2R}}(u-k_i)\left(\frac{|A(k_i,0,R)|}{R^n}\right)^{1+\frac{1}{p}}.$$

We now choose a number $\ell \in \mathbb{N}$ such that

$$2\tilde{c}_1\left[c_2\ell^{-\frac{n(p-1)}{(n-1)p}}\right]^{1+\frac{1}{p}} \leq 1$$

is satisfied (with c_2 denoting the constant from Lemma 3.11 for $\gamma = 1/2$). Hence, ℓ depends only on n, p, and C_0 and is in particular independent of the ball B_R. We now distinguish two cases:

(a) $\mathrm{osc}(0,2R) < 2^\ell R$: this is the trivial case, since the oscillations on B_R are bounded by a multiple of the radius R.
(b) $\mathrm{osc}(0,2R) \geq 2^\ell R$: with the choice of ℓ and Lemma 3.11, the previous L^∞-estimate gives

$$M(0,R/2) = \sup_{B_{R/2}} u \leq k_\ell + c_1(n,p,C_0)R + \frac{1}{2}(M(0,2R) - k_\ell).$$

Recalling the definition of k_ℓ and subtracting $m(0,R/2) \geq m(0,2R)$ on both sides of this inequality, we get

$$\mathrm{osc}(0,R/2) = M(0,R/2) - m(0,R/2)$$

$$\leq c_1(n,p,C_0)R + (1 - 2^{-\ell-2})\,\mathrm{osc}(0,2R).$$

In conclusion, we have in both cases the estimate

$$\mathrm{osc}(0,R/2) \leq (1 - 2^{-\ell-2})\,\mathrm{osc}(0,2R) + c(n,p,C_0)R$$

$$= 4^{-\alpha_0}\,\mathrm{osc}(0,2R) + c(n,p,C_0)R$$

with exponent $\alpha_0 := -\log_4(1 - 2^{-\ell-2}) = \alpha_0(n,p,C_0) > 0$. Therefore, the application of the iteration Lemma B.3 yields

$$\mathrm{osc}(0,r) \leq c(\alpha_0,\alpha)\left[\left(\frac{r}{R}\right)^\alpha \mathrm{osc}(0,2R) + c(n,p,C_0)r^\alpha\right]$$

for every exponent $\alpha \in (0, \alpha_0)$ and all $r \leq R$. Consequently, for every $\Omega' \Subset \Omega$, the α-Hölder norm $\|u\|_{C^{0,\alpha}(\Omega')}$ is bounded by a constant depending only on n, p, C_0, Ω and Ω' (which might diverge as $\mathrm{dist}(\Omega', \partial\Omega) \searrow 0$ since close to the boundary we can only work with very small balls). This concludes the proof of the theorem. $\qquad\square$

Via Lemma 3.4 and Corollary 3.5, the regularity result of Theorem 3.12 applies in particular to Q-minimizers if the integrand of the functional F satisfies the assumption of p-growth. Therefore, since both weak solutions to elliptic systems and minimizers to variational functionals are in fact Q-minimizers of such functionals under mild assumptions, see Examples 3.3, we get, as immediate consequences, the following Hölder regularity results.

Corollary 3.13 *Let $u \in W^{1,p}(\Omega)$, for some $p \in (1, \infty)$, be a weak solution to equation (3.1) with a vector field $a \colon \Omega \times \mathbb{R} \times \mathbb{R}^n \to \mathbb{R}^n$ and an inhomogeneity $a_0 \colon \Omega \times \mathbb{R} \times \mathbb{R}^n \to \mathbb{R}$ which satisfy the Carathéodory condition, the growth conditions (2.5) and (2.6), and such that $a(x, u, z) \cdot z \geq |z|^p$ holds for almost every $x \in \Omega$ and all $(u, z) \in \mathbb{R} \times \mathbb{R}^n$. Then there exists a positive exponent $\alpha = \alpha(n, p, L, \Omega, \|Du\|_{L^p(\Omega, \mathbb{R}^n)})$ (or $\alpha = \alpha(n, p, L)$ in the homogeneous case $a_0 = 0$) such that u is locally Hölder continuous in Ω with Hölder exponent α, i.e. $u \in C^{0,\alpha}(\Omega)$.*

Corollary 3.14 *Let $u \in W^{1,p}(\Omega)$, for some $p \in (1, \infty)$, be a minimizer of the functional (2.2) with an integrand $f \colon \Omega \times \mathbb{R} \times \mathbb{R}^n \to \mathbb{R}$ which satisfies the Carathéodory condition, the growth condition (2.7) and such that $f(x, u, z) \geq |z|^p$ holds for almost every $x \in \Omega$ and all $(u, z) \in \mathbb{R} \times \mathbb{R}^n$. Then there exists a positive exponent $\alpha = \alpha(n, p, L)$ such that u is locally Hölder continuous in Ω with Hölder exponent α, i.e. $u \in C^{0,\alpha}(\Omega)$.*

Remarks 3.15

(i) As already mentioned, De Giorgi [15] proved this Hölder regularity result in 1957, but it was obtained independently and simultaneously by Nash [69], and shortly after Moser [67] proposed a different strategy of proof (see the next section). Nowadays these regularity results are known as the De Giorgi–Nash–Moser theory, and for further extensions we refer to the monograph [53] of Ladyzhenskaya and Ural'tseva.

(ii) A direct application of De Giorgi's technique to minimizers of variational problems (this means not via the Euler–Lagrange equation, which not necessarily exists) was first given by Frehse [26] under stronger assumptions, and then by Giaquinta and Giusti [32] in full generality.

(iii) As the example below demonstrates, the result is sharp in the sense that we can expect only Hölder continuity for the solution for *some* exponent $\alpha \in (0, 1)$, but not for every exponent $\alpha \in (0, 1)$ (in particular, we cannot expect differentiability in the classical sense).

Example 3.16 *Consider $B_1 \subset \mathbb{R}^n$ with $n \geq 2$ and let $u \colon B_1 \to \mathbb{R}$ be given by $u(x) = x_1 |x|^{\alpha - 1}$ for some $\alpha \in (0, 1)$. Then $u \in W^{1,2}(B_1) \cap C^{0,\alpha}(\overline{B_1})$, $u \notin$*

$C^{0,\beta}(B_1)$ for $\beta > \alpha$, and u is a weak solution to the equation div $(a(x)Du) = 0$ in B_1, with measurable, bounded, elliptic coefficients $a\colon B_1 \to \mathbb{R}^{n\times n}$ defined by

$$a_{ij}(x) := \delta_{ij} + \frac{(1-\alpha)(n-1+\alpha)}{\alpha(n-2+\alpha)}\frac{x_i x_j}{|x|^2} \qquad \text{for } i,j \in \{1,\dots,n\}.$$

Proof The optimal Hölder continuity of u with exponent α is clear, and according to Example 1.43 (iii), u belongs to $W^{1,p}(B_1)$ for all $p \in [1, n/(1 - \alpha))$, so in particular to $W^{1,2}(B_1)$, with weak derivatives given by

$$D_j u(x) = \delta_{1j}|x|^{\alpha-1} + (\alpha - 1)x_1 x_j |x|^{\alpha-3}$$

for $j \in \{1,\dots,n\}$. Hence, it only remains to check that u is indeed a weak solution to the equation div $(a(x)Du) = 0$ in B_1. For $x \neq 0$ we easily calculate

$$\sum_{j=1}^{n} a_{ij}(x)D_j u(x) = \delta_{1i}|x|^{\alpha-1} + \frac{1-\alpha}{n-2+\alpha}x_1 x_i |x|^{\alpha-3},$$

which in turn implies

$$\sum_{i,j=1}^{n} D_i\big(a_{ij}(x)D_j u(x)\big)$$

$$= D_1|x|^{\alpha-1} + \frac{1-\alpha}{n-2+\alpha}\sum_{i=1}^{n} D_i\big(x_1 x_i |x|^{\alpha-3}\big)$$

$$= x_1|x|^{\alpha-3}\Big((\alpha-1) + \frac{1-\alpha}{n-2+\alpha}(1+n+\alpha-3)\Big) = 0.$$

The application of Lemma 1.41 then shows div $(a(x)Du) = 0$ in B_1 in the weak sense, since single points are negligible in the sense of the capacity condition (1.12) for $q = 2$, see Example 1.43 (i). $\qquad\square$

3.2   Moser's Iteration Technique

The aim of this section is to explain the iteration technique developed in Moser's paper [67], which allows for an alternative proof of the boundedness and regularity result for weak solutions to elliptic equations, which is a particular situation where De Giorgi's result from Theorem 3.9 and Corollary 3.10 applies. As in De Giorgi's approach, we proceed in two steps. In a first step, we show boundedness of weak solutions, more precisely, we here distinguish between boundedness from above for weak subsolutions and boundedness from below for weak supersolutions (which, in some sense, corresponds to boundedness from above for subminimizers and boundedness from below for superminimizers established in the previous section). In a second step, we then prove, relying on a Harnack-type inequality, Hölder regularity of weak solutions.

Before entering into the details of Moser's iteration technique, we first introduce the concept of weak subsolutions and weak supersolutions for quasilinear equations of the form (2.1) in the scalar case $N = 1$.

Definition 3.17 A function $u \in W^{1,p}_{\mathrm{loc}}(\Omega)$, for some $p \in [1,\infty)$, is called a *weak subsolution* (or *weak supersolution*) of the equation (2.1) if

$$x \mapsto a(x, u(x), Du(x)) \in L^{p'}_{\mathrm{loc}}(\Omega, \mathbb{R}^n),$$

$$x \mapsto a_0(x, u(x), Du(x)) \in L^{p'}_{\mathrm{loc}}(\Omega),$$

and if for all non-negative (non-positive) functions $\varphi \in C^\infty_0(\Omega)$ there holds

$$\int_\Omega a(x, u, Du) \cdot D\varphi \, dx \leq \int_\Omega a_0(x, u, Du)\varphi \, dx.$$

Remarks 3.18

(i) A function $u \in W^{1,p}_{\mathrm{loc}}(\Omega)$ is a weak solution in the sense of Definition 2.2 (with $q = p$) if and only if it is a weak subsolution and a weak supersolution.

(ii) Every subminimizer of a regular variational functional is a weak subsolution of the corresponding Euler–Lagrange equation (cf. Lemma 2.8), and reversely, every weak subsolution is sub-Q-minimizer of the functional $\tilde{E}_p$ given in Example 3.3 (ii), under the same assumptions as stated there.

In what follows, we restrict ourselves, for simplicity, to homogeneous, linear elliptic equations of the form

$$\mathrm{div}\,(a(x)Du) = 0 \qquad \text{in } \Omega, \tag{3.5}$$

with measurable coefficients $a \colon \Omega \to \mathbb{R}^{n \times n}$ which satisfy the following ellipticity and boundedness assumptions

$$a(x)\xi \cdot \xi \geq |\xi|^2 \tag{3.6}$$

$$a(x)\xi \cdot \tilde{\xi} \leq L|\xi||\tilde{\xi}| \tag{3.7}$$

for almost every $x \in \Omega$, all $\xi, \tilde{\xi} \in \mathbb{R}^n$ and some $L \geq 1$. In this case, an approximation argument shows that $u \in W^{1,2}(\Omega)$ is a weak subsolution if

$$\int_{\Omega} a(x) Du \cdot D\varphi \, dx \leq 0 \qquad (3.8)$$

holds for every non-negative function $\varphi \in W_0^{1,2}(\Omega)$ (and it is a weak supersolution if this integral inequality holds with the opposite sign). However, we note that the technique detailed below does not rely on the linear structure at all, and specifically Serrin [75] and Trudinger [81] have given extensions of the theory to more general quasilinear equations.

3.2.1   Local Boundedness

Moser's strategy for proving a local L^∞-estimate consists in showing, initially, that every weak solution actually belongs to the space L_{loc}^p for any $p \in [1, \infty)$ with a corresponding estimate. Subsequently, one then passes to the limit $p \to \infty$, which requires some sort of uniform boundedness of the involved constants. We follow the presentation in [5, Section 4] and [57, Chapter 2.3.5].

The central idea is to test the weak formulation of the elliptic equation (3.5) with *powers* of the weak solution. As a starting point, we find in this way a Caccioppoli-type inequality for powers of weak (sub- and super-) solutions.

Lemma 3.19 *Let $u \in W^{1,2}(\Omega)$ be a weak subsolution to the equation (3.5) with measurable coefficients $a: \Omega \to \mathbb{R}^{n \times n}$ satisfying (3.6) and (3.7), and let $t \geq 1$. Then we have*

$$(u_+)^t \in L_{\mathrm{loc}}^2(\Omega) \implies (u_+)^t \in W_{\mathrm{loc}}^{1,2}(\Omega).$$

Moreover, for every $s \geq 1$ and each $\eta \in C_0^\infty(\Omega, \mathbb{R}_0^+)$ the following Caccioppoli-type inequality holds true:

$$\int_{\Omega} \left| D\big((u_+)^t \eta^s\big) \right|^2 dx \leq 8L^2(t^2 + 1)s^2 \int_{\Omega} (u_+)^{2t} |D\eta|^2 \eta^{2s-2} \, dx . \qquad (3.9)$$

Proof We first observe that we may assume that u is non-negative (since otherwise, we pass to u_+, which is a non-negative subsolution of the same equation). One now wishes to use the function $u^{2t-1}\eta^{2s}$ for testing (3.8), but since it is not known to belong to the admissible function class $W_0^{1,2}(\Omega)$ for $t > 1$, we need to perform a truncation technique. To this end, for $K > 0$, we define the truncation operator $T_K: \mathbb{R} \to \mathbb{R}$ via $T_K y := \min\{y, K\}$. We then consider the non-negative function $\varphi := (T_K u)^{2t-2} u \eta^{2s}$ and compute

$$D\varphi = 2s(T_K u)^{2t-2} u D\eta \eta^{2s-1} + (T_K u)^{2t-2} Du \eta^{2s} + (2t-2) u^{2t-2} \mathbb{1}_{\{u \leq K\}} Du \eta^{2s}.$$

This shows that φ belongs to the space $W_0^{1,2}(\Omega)$, and φ can thus be used in the weak formulation (3.8) of subsolutions. Employing the ellipticity (3.6) and the boundedness (3.7) of the coefficients, we find

$$\int_\Omega (T_K u)^{2t-2} |Du|^2 \eta^{2s} \, dx \leq \int_\Omega (T_K u)^{2t-2} a(x) Du \cdot Du \eta^{2s} \, dx$$

$$\leq \int_\Omega \left[(T_K u)^{2t-2} + (2t-2) u^{2t-2} \mathbb{1}_{\{u \leq K\}} \right] a(x) Du \cdot Du \eta^{2s} \, dx$$

$$\leq -2s \int_\Omega (T_K u)^{2t-2} u\, a(x) Du \cdot D\eta \eta^{2s-1} \, dx$$

$$\leq 2Ls \int_\Omega (T_K u)^{2t-2} u |Du| |D\eta| \eta^{2s-1} \, dx$$

$$\leq \frac{1}{2} \int_\Omega (T_K u)^{2t-2} |Du|^2 \eta^{2s} \, dx + 2L^2 s^2 \int_\Omega u^{2t} |D\eta|^2 \eta^{2s-2} \, dx.$$

Absorbing the first integral on the right-hand side, we may pass to the limit $K \to \infty$ by Fatou's Theorem 1.9, and we obtain

$$\int_\Omega \eta^{2s} u^{2t-2} |Du|^2 \, dx \leq 4L^2 s^2 \int_\Omega u^{2t} |D\eta|^2 \eta^{2s-2} \, dx.$$

The assertion then follows from the inequality

$$|D(u^t \eta^s)| \leq t u^{t-1} |Du| \eta^s + s u^t |D\eta| \eta^{s-1}. \qquad \square$$

Similarly as in De Giorgi's approach, the previous Caccioppoli-type inequalities are the key estimates for proving boundedness of weak solutions. Therefore, one might introduce a *Moser class* of functions $v \in W^{1,2}(\Omega)$ for which there exists a constant M_0 such that for all $s, t \geq 1$ and every $\eta \in C_0^\infty(\Omega, \mathbb{R}_0^+)$ a Caccioppoli-type inequality of the form

$$\int_\Omega \left| D(|v|^t \eta^s) \right|^2 dx \leq M_0(t^2 + 1) s^2 \int_\Omega |v|^{2t} |D\eta|^2 \eta^{2s-2} \, dx. \qquad (3.10)$$

holds (however, notice that the validity for all $s \geq 1$ is no restriction, but imposed only for later convenience). By linearity of the equation (3.5) and with the previous lemma at hand, we immediately observe that for every weak supersolution u to (3.5) the function u_- satisfies such a condition, and, in turn, also every weak solution u.

Via Sobolev's inequality, we first improve the integrability by a factor $2^*/2$ (which would be interpreted as any arbitrary finite number in the simpler, therefore not explicitly stated two-dimensional case $n = 2$) and infer a reverse Hölder inequality for functions in the Moser class, cf. [67, Lemma 1]. Hence, similarly as in De Giorgi's approach, the only prerequisite for doing so is the validity of a suitable Caccioppoli inequality.

Lemma 3.20 (Moser) *Consider $v \in W^{1,2}(\Omega)$ and a pair of concentric balls $B_r(x_0) \Subset B_R(x_0) \subset \Omega$. If v satisfies the Caccioppoli-type inequality (3.10) for some $t \geq 1$, for $s := 1 + nt - n/2$ and every function $\eta \in C_0^\infty(B_R(x_0), [0,1])$ with $\eta \equiv 1$ on $B_r(x_0)$ and $|D\eta| \leq 2/(R-r)$, then we also have*

$$\left(\int_{B_R(x_0)} (|v|\eta^n)^{\frac{2n}{n-2}t} \eta^{-n} \, dx \right)^{\frac{n-2}{n}} \leq c(n) M_0 \frac{t^4}{(R-r)^2} \int_{B_R(x_0)} (|v|\eta^n)^{2t} \eta^{-n} \, dx .$$

Proof We first note $1 \leq s \leq nt$. Then, due to the choice of s, the claim is inferred from (3.9) by Sobolev's inequality from Lemma 1.51 as follows

$$\left(\int_{B_R(x_0)} (|v|\eta^n)^{\frac{2n}{n-2}t} \eta^{-n} \, dx \right)^{\frac{n-2}{n}}$$

$$= \left(\int_{B_R(x_0)} (|v|^t \eta^{nt-\frac{n}{2}+1})^{\frac{2n}{n-2}} \, dx \right)^{\frac{n-2}{n}}$$

$$\leq c(n) \int_{B_R(x_0)} \left| D(|v|^t \eta^{nt-\frac{n}{2}+1}) \right|^2 dx$$

$$\leq c(n) M_0 t^4 (R-r)^{-2} \int_{B_R(x_0)} |v|^{2t} \eta^{2nt-n} \, dx$$

$$= c(n) M_0 t^4 (R-r)^{-2} \int_{B_R(x_0)} (|v|\eta^n)^{2t} \eta^{-n} \, dx . \qquad \square$$

Since the integrals on the right-hand side and on the left-hand side of the inequality in Lemma 3.20 are of the same form, we can easily iterate these reverse Hölder inequalities. This yields implications of the type

$$v \in L^2(B_R(x_0)) \Rightarrow v \in L_{\text{loc}}^{2\frac{n}{n-2}}(B_R(x_0)) \Rightarrow \ldots \Rightarrow v \in L_{\text{loc}}^{2(\frac{n}{n-2})^j}(B_R(x_0))$$

for every $j \in \mathbb{N}$. Consequently, we obtain in particular that v belongs to $L^p(B_r(x_0))$, and hence, that every function in the Moser class belongs to $L_{\text{loc}}^p(\Omega)$, for every finite exponent $p < \infty$. Moreover, we even get local boundedness by carefully estimating all constants that are involved in the iteration.

Lemma 3.21 *Consider $v \in W^{1,2}(\Omega)$ and a pair of concentric balls $B_r(x_0) \Subset B_R(x_0) \subset \Omega$. If v satisfies, for some constant M_0, the Caccioppoli-type inequality (3.10) for all $t \geq 1$, all $s \geq 1$ and every $\eta \in C_0^\infty(B_R(x_0), [0,1])$ with $\eta \equiv 1$ on $B_r(x_0)$ and $|D\eta| \leq 2/(R-r)$, then v is bounded in $B_r(x_0)$ and satisfies the estimate*

$$\sup_{B_r(x_0)} |v|^2 \leq c(n, M_0)(R-r)^{-n} \int_{B_R(x_0)} |v|^2 \, dx.$$

Proof In order to iterate the inequality in Lemma 3.20 as announced, we introduce for $j \in \mathbb{N}_0$ the abbreviations

$$t_j := \left(\frac{n}{n-2}\right)^j,$$

$$\Psi_j := \left(\int_{B_R(x_0)} (|v|\eta^n)^{2t_j} \eta^{-n} \, dx\right)^{\frac{1}{t_j}},$$

$$A_j := \left(\frac{c(n)M_0 t_j^4}{(R-r)^2}\right)^{\frac{1}{t_j}},$$

where $c(n)$ denotes the constant from Lemma 3.20. With this terminology the estimate in Lemma 3.20 reads as

$$\Psi_{j+1} \leq A_j \Psi_j,$$

and by iterating this inequality we conclude $\Psi_{m+1} \leq (\prod_{j=0}^m A_j)\Psi_0$ for every $m \in \mathbb{N}$. Since the infinite product

$$\prod_{j=0}^\infty A_j = \prod_{j=0}^\infty \left(\frac{c(n)M_0 t_j^4}{(R-r)^2}\right)^{\frac{1}{t_j}}$$

$$= \left(\frac{c(n)M_0}{(R-r)^2}\right)^{\sum_{j=0}^\infty \left(\frac{n-2}{n}\right)^j} \left(\frac{n}{n-2}\right)^{4\sum_{j=0}^\infty j\left(\frac{n-2}{n}\right)^j}$$

$$= \left(\frac{c(n)M_0}{(R-r)^2}\right)^{\frac{n}{2}} \left(\frac{n}{n-2}\right)^{n(n-2)}$$

converges, we can pass to the limit $m \to \infty$. Employing $\eta \equiv 1$ in $B_r(x_0)$ and taking into account that $\|w\|_{L^\infty(\Omega)} = \lim_{p\to\infty} \|w\|_{L^p(\Omega)}$ holds for every measurable function w defined on Ω, we then obtain the assertion

$$\sup_{B_r(x_0)} |v|^2 \leq \lim_{m\to\infty} \Psi_m \leq c(n, M_0)(R-r)^{-n}\Psi_0$$

$$\leq c(n, M_0)(R-r)^{-n} \int_{B_R(x_0)} |v|^2 \, dx. \qquad \square$$

As a direct consequence of Lemma 3.19, we obtain the local boundedness from above for weak subsolutions to the equation (3.5), and, by linearity of the equation, local boundedness from below for weak supersolutions to the equation (3.5), cf. [67, Theorem 1]. This result was previously obtained via De Giorgi's approach in Sect. 3.1.1 for sub-Q-minimizers and super-Q-minimizers (which applies also to the special case of linear elliptic equations).

Theorem 3.22 (Moser) *Let $u \in W^{1,2}(\Omega)$ be a weak solution to the equation (3.5) with measurable coefficients $a \colon \Omega \to \mathbb{R}^{n \times n}$ satisfying (3.6) and (3.7). Then u is locally bounded, i.e. $u \in L^{\infty}_{\mathrm{loc}}(\Omega)$, and for every pair of concentric balls $B_r(x_0) \Subset B_R(x_0) \subset \Omega$ we have*

$$\sup_{B_r(x_0)} |u|^2 \leq c(n, L)(R - r)^{-n} \int_{B_R(x_0)} |u|^2 \, dx \,.$$

If u is only a weak subsolution or a weak supersolution, then the same inequality is true, with u replaced by u_+ and u_-, respectively.

For later convenience we further state another explicit estimate, in which arbitrary positive powers of the weak solution appear.

Corollary 3.23 *Let $u \in W^{1,2}(\Omega)$ be a weak solution to the equation (3.5) with measurable coefficients $a \colon \Omega \to \mathbb{R}^{n \times n}$ satisfying (3.6) and (3.7). Then, for every ball $B_R(x_0) \subset \Omega$ and every $q > 0$, there holds*

$$\sup_{B_{R/2}(x_0)} |u| \leq c(n, L, q)\left(\fint_{B_R(x_0)} |u|^q \, dx \right)^{\frac{1}{q}} \,.$$

Proof With the result of Theorem 3.22 for the choice $r = R/2$ at hand, the claim follows immediately for $q \geq 2$ via Jensen's inequality

$$\sup_{B_{R/2}(x_0)} |u| \leq c(n, L)\left(\fint_{B_R(x_0)} |u|^2 \, dx \right)^{\frac{1}{2}} \leq c(n, L)\left(\fint_{B_R(x_0)} |u|^q \, dx \right)^{\frac{1}{q}} \,.$$

If the case $q \in (0, 2)$ is considered instead, we initially observe via Young's inequality (1.3)

$$\phi(\varrho) := \sup_{B_\varrho(x_0)} |u| \leq c(n, L)\left((\sigma - \varrho)^{-n} \int_{B_\sigma(x_0)} |u|^q \, dx \right)^{\frac{1}{2}} \phi(\sigma)^{\frac{2-q}{2}}$$

$$\leq c(n, L)\left((\sigma - \varrho)^{-n} \int_{B_R(x_0)} |u|^q \, dx \right)^{\frac{1}{q}} + \frac{1}{2}\phi(\sigma)$$

for all $R/2 \leq \varrho < \sigma \leq R$, which in turn implies the assertion via the iteration Lemma B.1. $\qquad\qquad\square$

3.2.2 Local Hölder Continuity

Before addressing the local Hölder regularity of weak solutions to the linear equation (3.5), we first establish a Harnack inequality, which allows us to estimate the maximum of a positive function in terms of its infimum. To this end, we derive a quantitative lower bound for the infimum of a positive weak supersolution (which, due to its boundedness from below from the previous Theorem 3.22, may be supposed positive).

Lemma 3.24 *Let $u \in W^{1,2}(\Omega)$ be a positive weak supersolution to the equation (3.5) with measurable coefficients $a \colon \Omega \to \mathbb{R}^{n \times n}$ satisfying (3.6) and (3.7). Then, for every ball $B_R(x_0) \subset \Omega$ and every $q > 0$, there holds*

$$\inf_{B_{R/2}(x_0)} u \geq c(n, L, q) \Big(\fint_{B_R(x_0)} u^{-q} \, dx \Big)^{-\frac{1}{q}},$$

and we further have $D(\log u) \in L^{2, n-2}_{\mathrm{loc}}(\Omega, \mathbb{R}^n)$ with

$$\int_{B_{R/2}(x_0)} |D(\log u)|^2 \, dx \leq c(n, L) R^{n-2}.$$

Proof The first assertion is derived in a similar way as the statement of Lemma 3.19, with the crucial difference, that we now wish to test the equation with *negative powers* of the solution, instead of positive ones. To this end, for $K > 0$, we define the operator $M_K \colon \mathbb{R}^+ \to \mathbb{R}^+$ via $M_K y := y + K^{-1}$ and we then use the non-negative function $\varphi := (M_K u)^{-2t-1} \eta^{2s} \in W_0^{1,2}(\Omega)$, for $t \geq 0$, $s \geq 1$ and an arbitrary function $\eta \in C_0^\infty(\Omega, \mathbb{R}_0^+)$, in the weak formulation of supersolutions, which gives

$$\int_\Omega a(x) Du \cdot D\varphi \, dx \geq 0.$$

With the identity

$$D\varphi = 2s(M_K u)^{-2t-1} D\eta \eta^{2s-1} + (-2t - 1)(M_K u)^{-2t-2} Du \eta^{2s}$$

we then find, by taking advantage of (3.6) and (3.7), the estimate

$$(2t + 1) \int_\Omega (M_K u)^{-2t-2} |Du|^2 \eta^{2s} \, dx \leq 2Ls \int_\Omega (M_K u)^{-2t-1} |Du| |D\eta| \eta^{2s-1} \, dx$$

which, via Young's inequality, implies

$$\Big(t + \frac{1}{4} \Big) \int_\Omega (M_K u)^{-2t-2} |Du|^2 \eta^{2s} \, dx \leq L^2 s^2 \int_\Omega (M_K u)^{-2t} |D\eta|^2 \eta^{2s-2} \, dx.$$

$$\tag{3.11}$$

Now we distinguish the cases $t > 0$ and $t = 0$. In the first case, via the monotone convergence Theorem 1.10 and the estimate $|D(u^{-t}\eta^s)| \le tu^{-t-1}|Du|\eta^s + su^{-t}|D\eta|\eta^{s-1}$, we arrive at the Caccioppoli-type inequality

$$\int_\Omega |D(u^{-t}\eta^s)|^2 \, dx \le 2L^2(t+1)s^2 \int_\Omega u^{-2t}|D\eta|^2\eta^{2s-2} \, dx \,,$$

which, by arbitrariness of the localization function η, in turn, provides the implication

$$u^{-t} \in L^2_{\text{loc}}(\Omega) \implies u^{-t} \in W^{1,2}_{\text{loc}}(\Omega) \,.$$

In particular, for every $q > 0$, the function $u^{-q/2}$ satisfies the Caccioppoli-type inequality in (3.10), for all $s, t \ge 1$, every arbitrary function $\eta \in C^\infty_0(\Omega, \mathbb{R}^+_0)$ and some constant M_0 depending only on L and q. Consequently, we may apply Lemma 3.21 to the function $u^{-q/2}$ instead of v, to find

$$\sup_{B_{R/2}(x_0)} u^{-q} \le c(n, L, q) \fint_{B_R(x_0)} u^{-q} \, dx$$

for every ball $B_R(x_0) \subset \Omega$, and the proof of the first assertion is complete. In order to establish the second claim, we consider the case $t = 0$, $s = 1$ in (3.11) and derive

$$\int_\Omega |D(\log u)|^2\eta^2 \, dx \le 4L^2 \int_\Omega |D\eta|^2 \, dx \,.$$

The desired local Morrey space regularity of $D(\log u)$ follows if η is chosen in $C^\infty_0(B_R(x_0), [0,1])$ with $\eta \equiv 1$ on $B_{R/2}(x_0)$ and $|D\eta| \le 4/R$. $\qquad\square$

The last crucial ingredient of Moser's technique is the application of the John–Nirenberg lemma, which allows us to infer from the previous lemma a Harnack-type inequality, cf. [67, Theorem 2].

Lemma 3.25 (Moser's Harnack inequality) *Let $u \in W^{1,2}(\Omega)$ be a non-negative weak solution to the equation (3.5) with measurable coefficients $a\colon \Omega \to \mathbb{R}^{n\times n}$ satisfying (3.6) and (3.7). Then, for every ball $B_{\sqrt{n}R}(x_0) \Subset \Omega$, there holds*

$$\inf_{B_{R/2}(x_0)} u \ge c(n, L) \sup_{B_{R/2}(x_0)} u \,.$$

Proof We may suppose that u is positive (otherwise we prove the inequality for u replaced by $u + \varepsilon$, and the desired inequality then follows in the limit $\varepsilon \searrow 0$). Starting from the previous observation $D(\log u) \in L^{2,n-2}_{\text{loc}}(\Omega, \mathbb{R}^n)$, we initially observe $\log u \in \mathcal{L}^{2,n}_{\text{loc}}(\Omega)$. Via Theorem 1.26 of John–Nirenberg and the inclusions $B_R(x_0) \subset x_0 + (-R, R)^n \subset B_{\sqrt{n}R}(x_0)$, this implies

$$
\left(\fint_{B_R(x_0)} u^q \, dx \right) \left(\fint_{B_R(x_0)} u^{-q} \, dx \right)
$$

$$
= \left(\fint_{B_R(x_0)} \exp(q \log u) \, dx \right) \left(\fint_{B_R(x_0)} \exp(-q \log u) \, dx \right) \le c(n, L)
$$

for q sufficiently small depending only on n and L. Taking advantage of this fact, we find via Lemma 3.24 and Corollary 3.23

$$
\inf_{B_{R/2}(x_0)} u \ge c(n, L) \left(\fint_{B_R(x_0)} u^{-q} \, dx \right)^{-\frac{1}{q}}
$$

$$
\ge c(n, L) \left(\fint_{B_R(x_0)} u^q \, dx \right)^{\frac{1}{q}} \ge c(n, L) \sup_{B_{R/2}(x_0)} u. \qquad \square
$$

Finally, we infer the Hölder regularity of weak solutions as a consequence of Harnack's inequality and, hence, we complete the alternative proof of De Giorgi's regularity result from Corollary 3.13 for $W^{1,2}$-weak solutions to linear, elliptic equations.

Theorem 3.26 Let $u \in W^{1,2}(\Omega)$ be a weak solution to the equation (3.5) with measurable coefficients $a \colon \Omega \to \mathbb{R}^{n \times n}$ satisfying (3.6) and (3.7). Then there exists a positive exponent $\alpha = \alpha(n, L)$ such that u is locally Hölder continuous in Ω with Hölder exponent α, i.e. $u \in C^{0,\alpha}(\Omega)$.

Proof We here follow the presentation in [40, Chapter 7.9]. Due to Theorem 3.22, we already know that the supremum and infimum of u are locally bounded, and it hence remains to find an estimate for a suitable Hölder semi-norm. Using the notation for supremum, infimum and oscillation of u on balls in Ω as introduced at the beginning of Sect. 3.1.2, we now apply Harnack's inequality from Lemma 3.25 for a given ball $B_{\sqrt{n}R}(x_0) \Subset \Omega$ to the (non-negative) functions

$$
M(x_0, R) - u \quad \text{and} \quad u - m(x_0, R) \, .
$$

In this way, we find

$$
M(x_0, R) - m(x_0, R/2) \le c(n, L) \big(M(x_0, R) - M(x_0, R/2) \big) \, ,
$$

$$
M(x_0, R/2) - m(x_0, R) \le c(n, L) \big(m(x_0, R/2) - m(x_0, R) \big) \, ,
$$

with the same constants. Summing these inequalities up, we obtain

$$
\mathrm{osc}(x_0, R) + \mathrm{osc}(x_0, R/2) \le c(n, L) \big(\mathrm{osc}(x_0, R) - \mathrm{osc}(x_0, R/2) \big) \, ,
$$

and, hence, we have

$$\operatorname{osc}(x_0, R/2) \leq 2^{-\alpha} \operatorname{osc}(x_0, R) \quad \text{for } \alpha \in (0,1] \text{ satisfying } 2^{-\alpha} \geq \frac{c(n,L)-1}{c(n,L)+1}.$$

In particular, α is independent of the point x_0 under consideration. At this stage, we iterate this inequality and infer first $\operatorname{osc}(x_0, 2^{-j}R) \leq 2^{-j\alpha} \operatorname{osc}(x_0, R)$ for all $j \in \mathbb{N}$. For every $r \in (0, R]$ we then determine the unique number $j_0 \in \mathbb{N}_0$ such that $2^{-j_0-1}R < r \leq 2^{-j_0}R$, which finally yields the assertion via

$$\operatorname{osc}(x_0, r) \leq \operatorname{osc}(x_0, 2^{-j_0}R) \leq 2^{-j_0\alpha} \operatorname{osc}(x_0, R) \leq 2^{\alpha} \left(\frac{r}{R}\right)^{\alpha} \operatorname{osc}(x_0, R). \quad \square$$

Chapter 4
Foundations for the Vectorial Case

In this chapter we continue to investigate the regularity for weak solutions, but now we address the case of vector-valued solutions where we encounter fundamentally new phenomena when compared to the scalar case. In order to concentrate on the central concepts and ideas, we here restrict ourselves to the model case of quasilinear systems that are linear in the gradient variable, i.e., to systems of the form

$$\operatorname{div}\left(a(x,u)Du\right) = 0\,,$$

and we postpone the discussion of more general quasilinear elliptic systems in divergence form to Chap. 5. We first give two examples of elliptic systems, which admit a discontinuous or even unbounded weak solution. Then we investigate the optimal regularity of weak solutions in dependency of the "degree" of nonlinearity of the governing vector field a (under the permanent assumptions of suitable ellipticity, growth and regularity conditions on a). In this regard, we start by discussing the linear theory (that is for systems where the coefficients $a(x,u) \equiv a(x)$ do not depend explicitly on the weak solution) and establish full regularity estimates. This is quite peculiar and a consequence of the particular structure of the coefficients, since for more general systems, as in the counterexamples, one merely expects partial regularity results, that is, regularity outside of negligible sets. Secondly, we present three different strategies for proving partial $C^{0,\alpha}$-regularity results for such systems, where the coefficients may depend also explicitly on the weak solution. More precisely, we explain the main ideas for the blow-up technique, the method of $\mathcal{A}$-harmonic approximation, and the indirect approach.

© Springer International Publishing Switzerland 2016
L. Beck, *Elliptic Regularity Theory*, Lecture Notes of the Unione
Matematica Italiana 19, DOI 10.1007/978-3-319-27485-0_4

4.1   Counterexamples to Full Regularity

We now present the principal ideas for the construction of some counterexamples to full regularity, which were given first by De Giorgi [16] and shortly after by Giusti and Miranda [42]. However, we remark that similar examples, which exhibit the same features, were constructed independently by May'za [58]. The main intention, why we discuss the construction in some details, is to point out that the existence of discontinuous weak solutions is not a pathological phenomenon, which occurs only for very complicated and highly nonlinear elliptic systems. In fact, the examples presented here deal with discontinuous functions $u \in W^{1,2}(B_1, \mathbb{R}^n)$ solving the weak formulation for linear or quasilinear systems in divergence form of the type $\mathrm{div}\,(A(x)Du) = 0$ and $\mathrm{div}\,(\tilde{A}(u)Du) = 0$, respectively, in $B_1 \subset \mathbb{R}^n$ with $n \geq 3$. To give a good motivation for the choice of the coefficients, it is actually the most illustrative to start from one of the simplest functions in B_1 with a discontinuity, namely from the function

$$u(\alpha, x) := |x|^{-\alpha} x$$

for some $\alpha \in [1, n/2)$. Obviously, this function is discontinuous only at the origin and smooth everywhere else, and it also belongs to the Sobolev space $W^{1,2}(B_1, \mathbb{R}^n)$, as we have already observed in Example 1.43. For $x \neq 0$ its weak (and in fact classical) partial derivatives are given by the formula

$$D_i u^\kappa(\alpha, x) = |x|^{-\alpha} \delta_{i\kappa} - \alpha |x|^{-\alpha-2} x_i x_\kappa$$

for every $i, \kappa \in \{1, \ldots, n\}$. This very particular structure of the derivatives will now play the crucial role in the construction of the coefficients, and for later convenience, we observe already at this stage the following three identities:

$$\sum_{i=1}^n D_i u^i(\alpha, x) = \mathrm{Tr}(Du(\alpha, x)) = (n - \alpha)|x|^{-\alpha},$$

$$\sum_{i,\kappa=1}^n x_i x_\kappa D_i u^\kappa(\alpha, x) = (1 - \alpha)|x|^{2-\alpha},$$

$$\sum_{i,\kappa=1}^n (D_i u^\kappa(\alpha, x))^2 = |Du(\alpha, x)|^2 = (n - 2\alpha + \alpha^2)|x|^{-2\alpha}.$$

The counterexample of De Giorgi We first introduce a family of bilinear forms $A(b_1, b_2)$ on $\mathbb{R}^{n \times n}$ via

$$A_{ij}^{\kappa\lambda}(b_1, b_2, x) = \delta_{\kappa\lambda}\delta_{ij} + \left(b_1 \delta_{i\kappa} + b_2 \frac{x_i x_\kappa}{|x|^2}\right)\left(b_1 \delta_{j\lambda} + b_2 \frac{x_j x_\lambda}{|x|^2}\right)$$

for all indices $\kappa, \lambda, i, j \in \{1, \ldots, n\}$, $x \neq 0$, and with two arbitrary parameters $b_1, b_2 \in \mathbb{R}$, and we notice that the two constituent components $\delta_{i\kappa}$ and $x_i x_\kappa$ of the gradient $D_i u^\kappa$ reappear here. In what follows we shall use the convention

$$A(b_1, b_2, x)z \cdot \bar{z} = \sum_{\kappa, \lambda, i, j = 1}^{n} A_{ij}^{\kappa\lambda}(b_1, b_2, x)z_i^\kappa \bar{z}_j^\lambda$$

for all $z, \bar{z} \in R^{n \times n}$. From its definition we observe that $A(b_1, b_2, x)$ is bounded and elliptic with $A(b_1, b_2, x)z \cdot z \geq |z|^2$ for all $z \in R^{n \times n}$, all $x \neq 0$, and every choice of $b_1, b_2 \in \mathbb{R}$. It is further not difficult to check that the two free parameters b_1, b_2 can be chosen in such a way that, for each $\alpha \in [1, n/2)$, the function $u(\alpha)$ defined above is a weak solution to the associated linear system. To this end, we calculate, using the identities given above, for every $x \neq 0$ and all $\lambda, j \in \{1, \ldots, n\}$:

$$\big(A(b_1, b_2, x)Du(\alpha, x)\big)_j^\lambda$$

$$= D_j u^\lambda(\alpha, x)$$

$$+ \left(b_1 \sum_{i=1}^{n} D_i u^i(\alpha, x) + b_2 \sum_{i,\kappa=1}^{n} \frac{x_i x_\kappa}{|x|^2} D_i u^\kappa(\alpha, x)\right)\left(b_1 \delta_{j\lambda} + b_2 \frac{x_j x_\lambda}{|x|^2}\right)$$

$$= \big[b_1\big(b_1(n - \alpha) + b_2(1 - \alpha)\big) + 1\big]|x|^{-\alpha}\delta_{j\lambda}$$

$$+ \big[b_2\big(b_1(n - \alpha) + b_2(1 - \alpha)\big) - \alpha\big]|x|^{-\alpha-2}x_j x_\lambda.$$

Taking into account

$$\sum_{j=1}^{n} D_j(|x|^{-\alpha}\delta_{j\lambda}) = -\alpha|x|^{-\alpha-2}x_\lambda$$

and

$$\sum_{j=1}^{n} D_j(|x|^{-\alpha-2}x_j x_\lambda) = (n - 1 - \alpha)|x|^{-\alpha-2}x_\lambda,$$

we conclude that $\sum_{j=1}^{n} D_j(A(b_1, b_2, x)Du(\alpha, x))_j^\lambda$ vanishes for all $x \neq 0$ provided that α, b_1 and b_2 fulfill the equation

$$\alpha\big[b_1\big(b_1(n - \alpha) + b_2(1 - \alpha)\big) + 1\big] = (n - 1 - \alpha)\big[b_2\big(b_1(n - \alpha) + b_2(1 - \alpha)\big) - \alpha\big],$$

which can equivalently be written as

$$\alpha^2[(b_1 + b_2)^2 + 1] - \alpha n[(b_1 + b_2)^2 + 1] + (n - 1)b_2(b_2 + b_1 n) = 0.$$

In that case, as a direct application of Lemma 1.41, the function $u(\alpha)$ is in fact a weak solution to the system $\mathrm{div}\,(A(b_1,b_2,x)Du) = 0$ in B_1. Since, for each exponent α in the crucial interval $[1, n/2)$ of our interest, one can determine parameters $b_1, b_2 \in \mathbb{R}$ such that the previous equation is satisfied, we find on the one hand an example of an elliptic system that admits a bounded, discontinuous solution $(\alpha = 1)$ and on the other hand examples of elliptic systems that admit not only discontinuous, but even unbounded weak solutions $(\alpha \in (1, n/2))$. Specifically, with $b_1 = n - 2$, $b_2 = n$ and $\alpha \in (1, n/2)$ determined according to the previous identity, we have then obtained De Giorgi's counterexample from [16].

Example 4.1 (De Giorgi) *Assume $n \geq 3$ and let $u\colon \mathbb{R}^n \supset B_1 \to \mathbb{R}^n$ be given by*

$$u(\alpha, x) = |x|^{-\alpha} x \qquad for \; \alpha := \frac{n}{2}\big(1 - ((2n-2)^2 + 1)^{-1/2}\big).$$

Then $u \in W^{1,2}(B_1, \mathbb{R}^n)$ is an unbounded weak solution of the elliptic system

$$\mathrm{div}\,\big(A(n-2, n, x)Du(\alpha)\big) = 0 \qquad in \; B_1\,.$$

Remark 4.2 In the original formulation De Giorgi constructed the discontinuous functions $x \mapsto u(\alpha, x)$ in fact as minimizers of a quadratic variational functionals (defined by $w \mapsto \int_{B_1} A(n-2, n, x)Dw \cdot Dw\, dx$), and the equation above is the corresponding Euler–Lagrange system.

Since the coefficients are measurable, bounded and elliptic (hence, they satisfy all assumptions of the regularity theorems presented for the scalar case), De Giorgi's construction demonstrates that neither full Hölder regularity (for all weak solutions or even for all bounded weak solutions) nor local boundedness can in general be expected in the vectorial case $N > 1$. Consequently, an extension of Theorem 3.22 and Corollary 3.13 to the vectorial case is not possible for $n \geq 3$, and the same is true for the corresponding statements for minimizers of variational functionals.

The counterexample of Giusti and Miranda The coefficients in De Giorgi's example are discontinuous in the origin, and therefore, one might ask if irregular solutions to the system $\mathrm{div}\,(a(x, u)Du) = 0$ might also exist for coefficients $a(x, u)$ which are sufficiently regular in all variables. To answer this question, one needs to distinguish two cases. In the first case, if the coefficients depend only on the independent variable (that is, we have a linear system), then continuity or smoothness of the coefficients implies actually continuity or smoothness of every weak solution, as we will see in the next Sect. 4.2. Hence, a discontinuity in the coefficients turns out to be mandatory for the existence of a discontinuous weak solution in this case. Otherwise, if the coefficients are allowed to depend also explicitly on the weak solution, the question of the existence of an irregular weak solution was answered in the affirmative by Giusti and Miranda. By a modification of the above Example 4.1 they constructed an elliptic system whose coefficients depend smoothly on the weak solution and which admits a (bounded) discontinuous weak solution. To this end, we first observe that the function $u(1, x) = x/|x|$ is a weak solution to the system

$$\text{div}\,(A(1, 2/(n-2), x)Du) = 0 \qquad \text{in } B_1,$$

which corresponds to the choices $b_1 = 1$, $b_2 = 2/(n-2)$ and $\alpha = 1$. Then we may replace in $A(1, 2/(n-2), x)$ all occurrences of terms of the form $x_i/|x|$ by u_i. In view of $|u(x)| = 1$ for all $x \neq 0$, we obtain in this way the modified coefficients as

$$\tilde{A}_{ij}^{\kappa\lambda}(u) = \delta_{\kappa\lambda}\delta_{ij} + \left(\delta_{i\kappa} + \frac{4}{n-2}\frac{u_i u_\kappa}{1+|u|^2}\right)\left(\delta_{j\lambda} + \frac{4}{n-2}\frac{u_j u_\lambda}{1+|u|^2}\right),$$

for all indices $\kappa, \lambda, i, j \in \{1, \ldots, n\}$ and all $u \in \mathbb{R}^n$. Obviously, these coefficients are smooth in the u-variable, elliptic and bounded, and with these replacements we now end up with the counterexample of Giusti and Miranda from [42].

Example 4.3 (Giusti and Miranda) *Assume $n \geq 3$ and let $u \colon \mathbb{R}^n \supset B_1 \to \mathbb{R}^n$ be given by $u(x) = x/|x|$. Then $u \in W^{1,2}(B_1, \mathbb{R}^n) \cap L^\infty(B_1, \mathbb{R}^n)$, and u is a discontinuous weak solution of the elliptic system*

$$\text{div}\,(\tilde{A}(u)Du) = 0 \qquad \text{in } B_1. \tag{4.1}$$

Remarks 4.4

(i) The situation in the two-dimensional case $n = 2$ is different and all solutions are continuous (in fact, the gradient of every weak solution is as regular as the coefficients), see Theorem 5.31.

(ii) By introducing additional (dummy) variables, one can show that the function $u(x', x'') = x'/|x'|$ for $(x', x'') \in \mathbb{R}^3 \times \mathbb{R}^{n-3}$ with $n \geq 4$ is a bounded weak solution to a linear system of the form (4.1), and the set of points in which u is discontinuous has Hausdorff dimension $n - 3$.

Here in this case, with smooth dependence of the coefficients $\tilde{A}$ on all variables, the emergence of the discontinuity of the solution is obviously caused by the interaction between the solution and the gradient of the solution. Therefore, one might ask whether or not discontinuities or singularities are necessarily caused by some kind of interaction with the solution. As anticipated above, this is indeed the case if the system in linear is the gradient variable. However, even for analytic vector fields or integrands depending only on the gradient variable (but in a nonlinear way), there may exist non-smooth solutions, see the examples given in [46, 70, 71].

4.2 Linear Theory

We now provide some standard results and a priori estimates for weak solutions to linear systems of the form

$$\operatorname{div}\big(a(x)Du\big) = \operatorname{div} f - g \qquad \text{in } \Omega \tag{4.2}$$

with coefficients $a\colon \Omega \to \mathbb{R}^{Nn \times Nn}$, which are measurable, elliptic and bounded in the sense that

$$a(x)\xi \cdot \xi \geq |\xi|^2 \tag{4.3}$$

$$a(x)\xi \cdot \tilde{\xi} \leq L|\xi||\tilde{\xi}| \tag{4.4}$$

for almost every $x \in \Omega$, all $\xi, \tilde{\xi} \in \mathbb{R}^{Nn}$, and some $L \geq 1$. Furthermore, we always assume $f \in L^2(\Omega, \mathbb{R}^{Nn})$ and $g \in L^2(\Omega, \mathbb{R}^N)$, which essentially means that the right-hand side of (4.2) is a generic element of the dual space $(W_0^{1,2}(\Omega, \mathbb{R}^N))^*$. We shall see how higher differentiability properties of the data (that is, of the coefficients and the inhomogeneity) carry over to the gradient of the solution. Moreover, a similar effect occurs for decay properties of the solution, i.e., we also deal with Morrey- and Campanato-type regularity properties inherited by the data. Such results are very classical and usually referred to as "Schauder theory".

4.2.1 Hilbert Space Regularity

We first study Hilbert space regularity of weak solutions to linear systems of the form (4.2). This means that we start with a weak solution u in the Hilbert space $W^{1,2}(\Omega, \mathbb{R}^N)$ and we then look for optimal assumptions on the coefficients and the inhomogeneity which guarantee that u belongs to the Hilbert space $W_{\text{loc}}^{k,2}(\Omega, \mathbb{R}^N)$ for some $k \geq 1$, with a bound on the corresponding norm only in terms of the L^2-norm of u and of some suitable Sobolev norm of the data. For $k = 1$ such a bound is established via a Caccioppoli inequality. For later purposes we here state the Caccioppoli inequality for slightly more general elliptic systems of the form

$$\operatorname{div}\big(a(x, u)Du\big) = \operatorname{div} f - g \qquad \text{in } \Omega, \tag{4.5}$$

with f, g as above and with Carathéodory coefficients $a\colon \Omega \times \mathbb{R}^N \to \mathbb{R}^{Nn \times Nn}$ (now possibly depending on the u-variable), i.e., they are measurable with respect to x for all u and continuous in u for almost every x. Furthermore, we assume uniform ellipticity and boundedness in the sense that we have

$$a(x,u)\xi \cdot \xi \geq |\xi|^2 \tag{4.6}$$

$$a(x,u)\xi \cdot \tilde{\xi} \leq L|\xi||\tilde{\xi}| \tag{4.7}$$

for almost every $x \in \Omega$, all $u \in \mathbb{R}^N$, all $\xi, \tilde{\xi} \in \mathbb{R}^{Nn}$, and some $L \geq 1$.

Proposition 4.5 (Caccioppoli inequality) *Let* $u \in W^{1,2}(\Omega, \mathbb{R}^N)$ *be a weak solution to the system* (4.5) *with Carathéodory coefficients* $a \colon \Omega \times \mathbb{R}^N \to \mathbb{R}^{Nn \times Nn}$ *satisfying* (4.6) *and* (4.7), $f \in L^2(\Omega, \mathbb{R}^{Nn})$ *and* $g \in L^2(\Omega, \mathbb{R}^N)$. *Then we have for all* $\zeta \in \mathbb{R}^N$ *and all balls* $B_r(x_0) \Subset B_R(x_0) \subset \Omega$ *the estimate*

$$\int_{B_r(x_0)} |Du|^2 \, dx \leq c(L)(R-r)^{-2} \int_{B_R(x_0)} |u - \zeta|^2 \, dx$$

$$+ c \int_{B_R(x_0)} \left(|f|^2 + (R-r)^2 |g|^2 \right) dx.$$

Proof We take a cut-off function $\eta \in C_0^\infty(B_R(x_0), [0,1])$ which satisfies $\eta \equiv 1$ in $B_r(x_0)$ and $|D\eta| \leq 2(R-r)^{-1}$. We now test the weak formulation of the elliptic system (4.5) with the function $\varphi := \eta^2(u - \zeta) \in W_0^{1,2}(\Omega, \mathbb{R}^N)$. Thus, we get

$$\int_{B_R(x_0)} a(x,u)Du \cdot Du\eta^2 \, dx$$

$$= \int_{B_R(x_0)} a(x,u)Du \cdot \left[D(\eta^2(u - \zeta)) - 2\eta(u - \zeta) \otimes D\eta \right] dx$$

$$= \int_{B_R(x_0)} fD(\eta^2(u - \zeta)) \, dx + \int_{B_R(x_0)} g\eta^2(u - \zeta) \, dx$$

$$- 2 \int_{B_R(x_0)} a(x,u)Du \cdot ((u - \zeta) \otimes D\eta)\eta \, dx.$$

Using the ellipticity condition (4.6), the boundedness of a via (4.7) and Young's inequality, we next deduce

$$\int_{B_R(x_0)} |Du|^2 \eta^2 \, dx$$

$$\leq \frac{1}{2} \int_{B_R(x_0)} |Du|^2 \eta^2 \, dx + c(L) \int_{B_R(x_0)} |D\eta|^2 |u - \zeta|^2 \, dx$$

$$+ c \int_{B_R(x_0)} |f|^2 \, dx + \left| \int_{B_R(x_0)} g\eta^2(u - \zeta) \, dx \right| \tag{4.8}$$

$$\leq \frac{1}{2} \int_{B_R(x_0)} |Du|^2 \eta^2 \, dx + c(L)(R-r)^{-2} \int_{B_R(x_0)} |u - \zeta|^2 \, dx$$

$$+ c \int_{B_R(x_0)} \left(|f|^2 + (R-r)^2 |g|^2 \right) dx \,.$$

Absorbing the first integral on the right-hand side and employing the properties on η, we end up with the asserted inequality.   □

In the special situation that the weak solution happens to vanish on the boundary of $B_R(x_0)$, we have the following version of a Caccioppoli inequality, for which the domain of integration on both sides is equal to $B_R(x_0)$.

Proposition 4.6 *Let* $\Omega = B_R(x_0)$ *and consider a weak solution* $u \in W_0^{1,2}(B_R(x_0), \mathbb{R}^N)$ *to the system* (4.5) *with Carathéodory coefficients* $a \colon \Omega \times \mathbb{R}^N \to \mathbb{R}^{Nn \times Nn}$ *satisfying* (4.6) *and* (4.7), $f \in L^2(B_R(x_0), \mathbb{R}^{Nn})$ *and* $g \in L^2(B_R(x_0), \mathbb{R}^N)$. *Then we have*

$$\int_{B_R(x_0)} |Du|^2 \, dx \leq c(n, N) \int_{B_R(x_0)} \left(|f - (f)_{B_R(x_0)}|^2 + R^2 |g|^2 \right) dx \,.$$

Proof This statement is proved similarly to Proposition 4.5. We first note $\mathrm{div}\,(f)_{B_R(x_0)} = 0$, and hence, via testing the weak formulation of (4.5) with the function $\varphi := u \in W_0^{1,2}(B_R(x_0), \mathbb{R}^N)$ and Hölder's inequality we find

$$\int_{B_R(x_0)} |Du|^2 \, dx \leq \left(\int_{B_R(x_0)} |f - (f)_{B_R(x_0)}|^2 \, dx \right)^{\frac{1}{2}} \left(\int_{B_R(x_0)} |Du|^2 \, dx \right)^{\frac{1}{2}}$$

$$+ \left(\int_{B_R(x_0)} |g|^2 \, dx \right)^{\frac{1}{2}} \left(\int_{B_R(x_0)} |u|^2 \, dx \right)^{\frac{1}{2}} \,.$$

The assertion then follows from Poincaré's inequality.   □

In order to illustrate the importance of Caccioppoli inequalities and also the hole-filling technique, we briefly comment on weak solutions on the whole space and how to derive the well-known Liouville property of bounded weak solutions.

Corollary 4.7 *Let* $u \in W_{\mathrm{loc}}^{1,2}(\mathbb{R}^n, \mathbb{R}^N)$ *be a weak solution to*

$$\mathrm{div}\,(a(x, u)Du) = 0 \qquad \text{in } \mathbb{R}^n$$

with Carathéodory coefficients $a \colon \Omega \times \mathbb{R}^N \to \mathbb{R}^{Nn \times Nn}$ *satisfying* (4.6) *and* (4.7). *If* $Du \in L^2(\mathbb{R}^n, \mathbb{R}^{Nn})$ *holds, then* u *is constant in* $\mathbb{R}^n$. *In particular, every bounded weak solution is constant for* $n = 2$.

Proof Let $R > 0$ be arbitrary. From inequality (4.8) (with the choices $r = R/2$ and $\zeta = (u)_{B_R \setminus B_{R/2}}$) and Poincaré's inequality (for which the dependence on the radius R is again inferred by a rescaling argument), we first deduce

$$\int_{B_{R/2}} |Du|^2 \, dx \le c(L) R^{-2} \int_{B_R \setminus B_{R/2}} |u - (u)_{B_R \setminus B_{R/2}}|^2 \, dx \qquad (4.9)$$

$$\le c^*(n, N, L) \int_{B_R \setminus B_{R/2}} |Du|^2 \, dx \, .$$

Thus, adding $c^* \int_{B_{R/2}} |Du|^2 \, dx$ to both sides (that is, applying the hole-filling argument), we find

$$\int_{B_{R/2}} |Du|^2 \, dx \le \frac{c^*(n, N, L)}{c^*(n, N, L) + 1} \int_{B_R} |Du|^2 \, dx \, .$$

Since the constant on the left-hand side is independent of R and $Du \in L^2(\mathbb{R}^n, \mathbb{R}^{Nn})$ holds by assumption, we can pass to the limit $R \to \infty$, which gives

$$\int_{\mathbb{R}^n} |Du|^2 \, dx \le \frac{c^*(n, N, L)}{c^*(n, N, L) + 1} \int_{\mathbb{R}^n} |Du|^2 \, dx \, .$$

Since the constant on the right-hand side is strictly less than 1, we arrive at $Du \equiv 0$ in $\mathbb{R}^n$, and thus, u is constant in $\mathbb{R}^n$ as claimed.

Finally, if $u \in W^{1,2}_{\text{loc}}(\mathbb{R}^2, \mathbb{R}^N)$ is a bounded weak solution in the two-dimensional case, then we obtain from (4.9)

$$\int_{B_{R/2}} |Du|^2 \, dx \le c(n, L) R^{-2} R^2 \|u\|_{L^\infty(\mathbb{R}^2, \mathbb{R}^N)} \le c(n, L) \|u\|_{L^\infty(\mathbb{R}^2, \mathbb{R}^N)} \, .$$

Since R was arbitrary, this implies immediately $Du \in L^2(\mathbb{R}^2, \mathbb{R}^N)$, and constancy of u follows from the first claim. $\qquad \square$

Next, we return to the linear system (4.2) with coefficients $a(x)$ not depending explicitly on u and prove interior $W^{2,2}$-regularity for weak solutions. This can be considered as the toy case for the application of the difference quotient technique in order to obtain higher regularity. We will see later that this technique is also useful for more general vector fields $a(x, u, z)$ which are possibly nonlinear in the gradient variable (still under suitable assumptions concerning the differentiability in all variables, growth and ellipticity of the bilinear form $D_z a(x, u, z)$).

Proposition 4.8 *Consider a weak solution $u \in W^{1,2}(\Omega, \mathbb{R}^N)$ to the system (4.2) with coefficients $a \in W^{1,\infty}(\Omega, \mathbb{R}^{Nn \times Nn})$ satisfying (4.3), $f \in$*

$W^{1,2}(\Omega, \mathbb{R}^{Nn})$ and $g \in L^2(\Omega, \mathbb{R}^N)$. Then $u \in W^{2,2}_{\text{loc}}(\Omega, \mathbb{R}^N)$, and for all balls $B_r(x_0) \Subset B_R(x_0) \subset \Omega$ we have

$$\int_{B_r(x_0)} |D^2 u|^2 \, dx \leq c(n, \|a\|_{W^{1,\infty}(\Omega)}) [(R - r)^{-2} + 1] \int_{B_R(x_0)} |Du|^2 \, dx$$

$$+ c(n) \int_{B_R(x_0)} \left(|Df|^2 + |g|^2 \right) dx \, .$$

Proof We take $\varphi \in C_0^1(\Omega, \mathbb{R}^N)$ and test the weak formulation of (4.2) with the difference quotient $\triangle_{s,-h}\varphi$ of stepsize $-h \in \mathbb{R}$ such that $|h| \in (0, \text{dist}(\text{spt}\,\varphi, \partial\Omega))$ and in an arbitrary direction $s \in \{1, \ldots, n\}$. Using the integration by parts formula for finite difference quotients from Remark 1.45 (iii), we obtain

$$\int_\Omega \triangle_{s,h}(aDu) \cdot D\varphi \, dx = \int_\Omega \triangle_{s,h} f D\varphi \, dx + \int_\Omega \triangle_{s,h} g\varphi \, dx \, .$$

With the product formula for finite difference quotients we hence find that the function $v := \triangle_{s,h} u$ is a weak solution to the system

$$\text{div}\left(a(x)Dv(x) \right) = \text{div}\left(- \triangle_{s,h} a(x)Du(x + he_s) + \triangle_{s,h} f(x) \right) - \triangle_{s,h} g(x)$$

$$=: \text{div}\, \tilde{f}(x) - \triangle_{s,h} g(x)$$

in any subset $\Omega' \Subset \Omega$, provided that we suppose $|h| < \text{dist}(\Omega', \partial\Omega)$. At this stage the application of Proposition 4.5 would not yield that the family $\triangle_{s,h} Du$ is bounded in L^2, uniformly with respect to h, since g is assumed to belong only to $L^2(\Omega, \mathbb{R}^N)$ and consequently $\triangle_{s,h} g$ does not necessarily remain bounded. Nevertheless, the existence of second order derivatives of u can be shown by a similar argument.

Heuristically, we should think of all difference quotients $\triangle_{s,h}$ in the above derivation being replaced by the weak derivative D_s (which is rigorously not allowed). Hence, the function $w := D_s u$ would formally be a weak solution to the system

$$\text{div}\left(a(x)Dw \right) = \text{div}\left(- D_s a(x)Du + D_s f \right) - D_s g \, . \tag{4.10}$$

At this stage, with $-D_s a(x)Du + D_s f - ge_s$ and the zero-function playing the roles of f and g in Proposition 4.5, the desired L^2-estimate for Dw would follow immediately.

In a rigorous way, the problem that we need to work with difference quotients instead of weak derivatives and that consequently difference quotients of g appear is resolved as follows: we go back to the proof of Proposition 4.5 and estimate the integral involving g explicitly. For this purpose, given an open subset $\Omega' \Subset \Omega$, we consider concentric balls $B_r(x_0) \Subset B_R(x_0) \subset \Omega'$, $h \in \mathbb{R} \setminus \{0\}$ with $|h| < \min\{\mathrm{dist}(\Omega', \partial\Omega), (R-r)/2\}$ and a cut-off function $\eta \in C_0^\infty(B_{(R+r)/2}(x_0), [0,1])$ satisfying $\eta \equiv 1$ in $B_r(x_0)$ and $|D\eta| \le 4/(R-r)$. With the integration by parts formula for finite difference quotients, Young's inequality and Lemma 1.46 we first observe that

$$\left| \int_{B_{(R+r)/2}(x_0)} \triangle_{s,h} g \eta^2 \triangle_{s,h} u \, dx \right| = \left| \int_{B_{(R+r)/2}(x_0)} g \triangle_{s,-h}(\eta^2 \triangle_{s,h} u) \, dx \right|$$

$$\le 2 \int_{B_{(R+r)/2}(x_0)} |g|^2 \, dx + \frac{1}{8} \int_{B_{(R+r)/2}(x_0)} |D(\eta^2 \triangle_{s,h} u)|^2 \, dx$$

$$\le 2 \int_{B_{(R+r)/2}(x_0)} \left(|g|^2 + |\triangle_{s,h} u|^2 |D\eta|^2 \right) dx$$

$$+ \frac{1}{4} \int_{B_{(R+r)/2}(x_0)} |\triangle_{s,h} Du|^2 \eta^2 \, dx .$$

From inequality (4.8) (for $\zeta = 0$), with $\triangle_{s,h} u$, $\tilde{f}$, $\triangle_{s,h} g$, and $(R+r)/2$ instead of u, f, g, and, respectively, R we therefore deduce

$$\int_{B_{(R+r)/2}(x_0)} |\triangle_{s,h} Du|^2 \eta^2 \, dx$$

$$\le \frac{1}{2} \int_{B_{(R+r)/2}(x_0)} |\triangle_{s,h} Du|^2 \eta^2 \, dx$$

$$+ c(\|a\|_{L^\infty(\Omega)}) \int_{B_{(R+r)/2}(x_0)} |D\eta|^2 |\triangle_{s,h} u|^2 \, dx$$

$$+ c \int_{B_R(x_0)} |\tilde{f}|^2 \, dx + \left| \int_{B_R(x_0)} \triangle_{s,h} g \eta^2 \triangle_{s,h} u \, dx \right|$$

$$\le \frac{3}{4} \int_{B_{(R+r)/2}(x_0)} |\triangle_{s,h} Du|^2 \eta^2 \, dx$$

$$+ c(\|a\|_{L^\infty(\Omega)}) \int_{B_{(R+r)/2}(x_0)} |D\eta|^2 |\triangle_{s,h} u|^2 \, dx$$

$$+ c\|a\|_{W^{1,\infty}(\Omega)}^2 \int_{B_{(R+r)/2}(x_0)} |Du(x + he_s)|^2 \, dx$$

$$+ c \int_{B_{(R+r)/2}(x_0)} \left(|\triangle_{s,h} f|^2 + |g|^2 \right) dx$$

$$\le \frac{3}{4} \int_{B_{(R+r)/2}(x_0)} |\triangle_{s,h} Du|^2 \eta^2 \, dx$$

$$+ c(\|a\|_{W^{1,\infty}(\Omega)})[(R-r)^{-2}+1] \int_{B_R(x_0)} |Du|^2 \, dx$$

$$+ c \int_{B_R(x_0)} (|D_s f|^2 + |g|^2) \, dx \, ,$$

where we once again have used Lemma 1.46. According to Lemma 1.48, we hence obtain the existence of the second order derivative $D_s Du \in L^2(B_r(x_0), \mathbb{R}^{Nn})$, and by arbitrariness of $B_r(x_0)$ we have established $D_s Du \in L^2_{\mathrm{loc}}(\Omega, \mathbb{R}^{Nn})$. Moreover, we have made the formal computation leading to the differentiated system (4.10) rigorous, and $D_s u$ is indeed a weak solution for every $\Omega' \Subset \Omega$. Passing to the limit $h \to 0$ in the previous estimate (as mentioned above, one could alternatively use the validity of a Caccioppoli inequality for the differentiated system equation), we then find the estimate

$$\int_{B_r(x_0)} |D_s Du|^2 \, dx \le c(\|a\|_{W^{1,\infty}(\Omega)})[(R-r)^{-2}+1] \int_{B_R(x_0)} |Du|^2 \, dx$$

$$+ c \int_{B_R(x_0)} (|D_s f|^2 + |g|^2) \, dx$$

(and $B_R(x_0) \subset \Omega$ is indeed allowed now). The summation over $s \in \{1, \ldots, n\}$ then completes the proof of the proposition. $\square$

By an iteration of Proposition 4.8 (and combined with the Caccioppoli inequality from Proposition 4.5) we obtain the general result on interior $W^{k,2}$-regularity.

Theorem 4.9 *Let $k \in \mathbb{N}$ and consider a weak solution $u \in W^{1,2}(\Omega, \mathbb{R}^N)$ to the system (4.2) with coefficients $a \in W^{k,\infty}(\Omega, \mathbb{R}^{Nn \times Nn})$ satisfying (4.3), $f \in W^{k,2}(\Omega, \mathbb{R}^{Nn})$ and $g \in W^{k-1,2}(\Omega, \mathbb{R}^N)$. Then $u \in W^{k+1,2}_{\mathrm{loc}}(\Omega, \mathbb{R}^N)$, and for all $\Omega' \Subset \Omega$ we have*

$$\|u\|_{W^{k+1,2}(\Omega', \mathbb{R}^{Nn^{k+1}})} \le c(\|u\|_{L^2(\Omega, \mathbb{R}^N)} + \|f\|_{W^{k,2}(\Omega, \mathbb{R}^{Nn})} + \|g\|_{W^{k-1,2}(\Omega, \mathbb{R}^N)})$$

for a constant c depending only on n, k, $\|a\|_{W^{k,\infty}(\Omega)}$, Ω', and $\mathrm{dist}(\Omega', \partial\Omega)$.

As a direct consequence of the embedding result in Corollary 1.64 for Sobolev functions we finally conclude that every weak solution is smooth in the interior, provided that the coefficients and the inhomogeneity is smooth.

Corollary 4.10 *Consider a weak solution $u \in W^{1,2}(\Omega, \mathbb{R}^N)$ to the system (4.2) with coefficients $a \in C^\infty(\Omega, \mathbb{R}^{Nn \times Nn})$ satisfying (4.3), $f \in C^\infty(\Omega, \mathbb{R}^{Nn})$ and $g \in C^\infty(\Omega, \mathbb{R}^N)$. Then we have $u \in C^\infty(\Omega, \mathbb{R}^N)$.*

4.2.2 Decay Estimates

We next establish some decay estimates for weak solutions to the linear system (4.2), in the sense that we provide the optimal scaling of the L^2-norm (of the weak solution and of their gradients, respectively) on small balls in terms of the radius. In doing so we give Morrey-type estimates in $L^{2,\lambda}$ and Campanato-type estimates in $\mathcal{L}^{2,\lambda}$, for the optimal parameter λ. Such decay estimates were first investigated in this form by Campanato in [9]. We follow his original strategy of proof and proceed in three steps. We first study the particular situation of homogeneous systems with constant, elliptic coefficients, i.e., of systems of the form

$$\operatorname{div}\left(aDu\right) = 0 \quad \text{in } \Omega. \tag{4.11}$$

Via a perturbation argument, this situation is then generalized in a second step to the inhomogeneous case with constant, elliptic coefficients and finally in a third step to general systems with continuous, elliptic coefficients.

Homogeneous systems with constant coefficients Due to Theorem 4.9 (or Corollary 4.10) we already know that every weak solution to the system (4.11) is smooth, and we now derive the optimal decay behavior of the L^2-norm of the weak solution on balls in terms of a suitable power of the radius.

Lemma 4.11 (Decay estimates I; Campanato) *Let $u \in W^{1,2}(\Omega, \mathbb{R}^N)$ be a weak solution to the system (4.11) with constant coefficients $a \in \mathbb{R}^{Nn \times Nn}$ satisfying (4.3) and (4.4). Then for all balls $B_r(x_0) \subset B_R(x_0) \subset \Omega$ we have*

$$\int_{B_r(x_0)} |u|^2 \, dx \le c \Big(\frac{r}{R}\Big)^n \int_{B_R(x_0)} |u|^2 \, dx$$

and

$$\int_{B_r(x_0)} |u - (u)_{B_r(x_0)}|^2 \, dx \le c \Big(\frac{r}{R}\Big)^{n+2} \int_{B_R(x_0)} |u - (u)_{B_R(x_0)}|^2 \, dx,$$

with constants c depending only on n, N, and L. Moreover, the same estimates are true if u is replaced by any derivative $D^k u$ for $k \in \mathbb{N}$.

Proof We may assume $r \le R/2$ (otherwise both inequalities are satisfied trivially with constants $c = 2^n$ and $c = 2^{n+2}$, respectively). Due to Theorem 4.9, we have $u \in W^{k,2}_{\mathrm{loc}}(\Omega, \mathbb{R}^N)$ for every $k \in \mathbb{N}$, and we further recall that via Sobolev's and Morrey's embedding from Corollary 1.64 we have

$$\|u\|_{L^\infty(B_{R/2}(x_0),\mathbb{R}^N)} \leq c(n,N,k,R)\|u\|_{W^{k,2}(B_{R/2}(x_0),\mathbb{R}^N)}$$

whenever $k > n/2$. Choosing for example $k = n$, we then obtain

$$\int_{B_r(x_0)} |u|^2\,dx \leq c(n)r^n\|u\|^2_{L^\infty(B_{R/2}(x_0),\mathbb{R}^N)}$$

$$\leq c(n,N,R)r^n\|u\|^2_{W^{n,2}(B_{R/2}(x_0),\mathbb{R}^N)}$$

$$\leq c(n,N,L,R)r^n\int_{B_R(x_0)} |u|^2\,dx\,,$$

where for the last inequality we have exploited the estimate given in Theorem 4.9. By a simple rescaling argument, considering the function $v(y) := u(Ry + x_0)$ on the unit ball B_1, we determine the dependence of the constant on the radius R and find $c(n,N,L,R) = c(n,N,L)R^{-n}$. This completes the proof of the first inequality.

To prove the second inequality, we first observe that also $D_s u$ is a weak solution to the same system (4.11), for every $s \in \{1,\dots,n\}$. Consequently, the first inequality, applied for Du instead of u, yields

$$\int_{B_r(x_0)} |Du|^2\,dx \leq c(n,N,L)\left(\frac{r}{R}\right)^n \int_{B_{R/2}(x_0)} |Du|^2\,dx\,.$$

With Poincaré's inequality from Lemma 1.56 and the Caccioppoli inequality from Proposition 4.5 (applied with $\zeta = (u)_{B_R(x_0)} \in \mathbb{R}^N$), the assertion then follows from the chain of inequalities

$$\int_{B_r(x_0)} |u - (u)_{B_r(x_0)}|^2\,dx \leq c(n,N)r^2 \int_{B_r(x_0)} |Du|^2\,dx$$

$$\leq c(n,N,L)r^2\left(\frac{r}{R}\right)^n \int_{B_{R/2}(x_0)} |Du|^2\,dx$$

$$\leq c(n,N,L)\left(\frac{r}{R}\right)^{n+2} \int_{B_R(x_0)} |u - (u)_{B_R(x_0)}|^2\,dx\,.$$

Finally, we observe that with u also each partial derivative $D^\beta u$ of order $|\beta| = k$ (for any $k \in \mathbb{N}$) is a solution to the same homogeneous, linear system with constant coefficients. Consequently, the same decay estimates hold true for arbitrary derivatives of u instead of u.   □

Inhomogeneous systems with constant coefficients We next derive similar decay estimates for weak solutions of inhomogeneous systems (4.2), still with constant, elliptic coefficients. The inhomogeneity is here viewed as a perturbation of the homogeneous situation, and the decay estimates essentially follow from the previous Lemma 4.11, by controlling the L^2-distance between (the gradients of) the solution of the inhomogeneous system and the solution of the homogeneous system (with a suitable boundary constraint), in terms of the difference of these two system, that is, of the inhomogeneity. This is the first example of a *comparison principle*, and for its implementation we need to guarantee the existence of the comparison function. To this end, we recall the following consequence of the Lax–Milgram Theorem A.11 concerning the existence of weak solutions in the Hilbert space $W^{1,2}$ with prescribed boundary values.

Remark 4.12 For linear systems of the form (4.2) with measurable coefficients $a\colon \Omega \to \mathbb{R}^{Nn \times Nn}$ satisfying (4.3) and (4.4) the Lax–Milgram Theorem A.11 ensures the existence of a (unique) solution $u \in u_0 + W_0^{1,2}(\Omega, \mathbb{R}^N)$ with prescribed boundary values $u_0 \in W^{1,2}(\Omega, \mathbb{R}^N)$. To verify this assertion, one applies Theorem A.11 in the Hilbert space $W_0^{1,2}(\Omega, \mathbb{R}^N)$, with bilinear form $B\colon W_0^{1,2}(\Omega, \mathbb{R}^N) \times W_0^{1,2}(\Omega, \mathbb{R}^N) \to \mathbb{R}$ and right-hand side $F \in (W_0^{1,2}(\Omega, \mathbb{R}^N))^*$ given by

$$B(v, w) := \int_\Omega a(x) Dv \cdot Dw \, dx,$$

$$F(v) := \int_\Omega \left(-a(x) Du_0 \cdot Dv + f \cdot Dv + g \cdot v \right) dx,$$

for all $v, w \in W_0^{1,2}(\Omega, \mathbb{R}^N)$. Then $u := u_0 + \Lambda(F) \in u_0 + W_0^{1,2}(\Omega, \mathbb{R}^N)$ is a weak solution to (4.2).

With this existence result at hand, we can now prove the decay estimates in the inhomogeneous case via the aforementioned comparison technique.

Lemma 4.13 (Decay estimates II; Campanato) *Let* $u \in W^{1,2}(\Omega, \mathbb{R}^N)$ *be a weak solution to the system* (4.2) *with constant coefficients* $a \in \mathbb{R}^{Nn \times Nn}$ *satisfying* (4.3) *and* (4.4), $f \in L^2(\Omega, \mathbb{R}^{Nn})$ *and* $g \in L^2(\Omega, \mathbb{R}^N)$. *Then for all balls* $B_r(x_0) \subset B_R(x_0) \subset \Omega$ *we have*

$$\int_{B_r(x_0)} |Du|^2 \, dx \leq c \Big[\Big(\frac{r}{R} \Big)^n \int_{B_R(x_0)} |Du|^2 \, dx$$

$$+ \int_{B_R(x_0)} \left(|f|^2 + R^2 |g|^2 \right) dx \Big]$$

and

$$\int_{B_r(x_0)} |Du - (Du)_{B_r(x_0)}|^2 \, dx$$

$$\leq c \Bigg[\Big(\frac{r}{R}\Big)^{n+2} \int_{B_R(x_0)} |Du - (Du)_{B_R(x_0)}|^2 \, dx$$

$$+ \int_{B_R(x_0)} \big(|f - (f)_{B_R(x_0)}|^2 + R^2 |g|^2\big) \, dx \Bigg],$$

with constants c depending only on n, N, and L.

Proof We fix $B_R(x_0) \subset \Omega$. According to Remark 4.12 we may write the weak solution as $u = v + w$, where $v \in u + W_0^{1,2}(B_R(x_0), \mathbb{R}^N)$ is the weak solution to

$$\operatorname{div}(aDv) = 0 \qquad \text{in } B_R(x_0),$$

and $w \in W_0^{1,2}(B_R(x_0), \mathbb{R}^N)$ is the weak solution to

$$\operatorname{div}(aDw) = \operatorname{div} f - g \qquad \text{in } B_R(x_0)$$

(notice that $v = u$ and $w = 0$ if both f and g vanish). For the function Dv we have the decay estimates provided by Lemma 4.11 at our disposal, whereas the function Dw is controlled in L^2 by the L^2-norms of f and g only, due to Proposition 4.6. Using twice the fact that for every function $\varphi \in L^2(\Omega, \mathbb{R}^N)$ the map $\zeta \mapsto \int_\Omega |\varphi - \zeta|^2 \, dx$ is minimized by the mean value $\zeta = (\varphi)_\Omega$, we hence find

$$\int_{B_r(x_0)} |Du - (Du)_{B_r(x_0)}|^2 \, dx \leq \int_{B_r(x_0)} |Du - (Dv)_{B_r(x_0)}|^2 \, dx$$

$$\leq 2 \int_{B_r(x_0)} |Dv - (Dv)_{B_r(x_0)}|^2 \, dx + 2 \int_{B_R(x_0)} |Dw|^2 \, dx$$

$$\leq c(n,N,L) \Big(\frac{r}{R}\Big)^{n+2} \int_{B_R(x_0)} |Dv - (Dv)_{B_R(x_0)}|^2 \, dx + 2 \int_{B_R(x_0)} |Dw|^2 \, dx$$

$$\leq c(n,N,L) \Bigg[\Big(\frac{r}{R}\Big)^{n+2} \int_{B_R(x_0)} |Du - (Du)_{B_R(x_0)}|^2 \, dx + \int_{B_R(x_0)} |Dw|^2 \, dx \Bigg]$$

$$\leq c(n,N,L) \Bigg[\Big(\frac{r}{R}\Big)^{n+2} \int_{B_R(x_0)} |Du - (Du)_{B_R(x_0)}|^2 \, dx$$

$$+ \int_{B_R(x_0)} \big(|f - (f)_{B_R(x_0)}|^2 + R^2 |g|^2\big) \, dx \Bigg].$$

This finishes the proof of the second inequality, and the first one is obtained similarly (and in a simpler way since the passage between different mean values of Du is not needed). □

As a direct consequence of Lemma 4.13, combined with the iteration Lemma B.3, we thus obtain the following Morrey- and Campanato-space regularity result.

Corollary 4.14 *Let* $u \in W^{1,2}(\Omega, \mathbb{R}^N)$ *be a weak solution to the system* (4.2) *with constant coefficients* $a \in \mathbb{R}^{Nn \times Nn}$ *satisfying* (4.3) *and* (4.4), $f \in L^2(\Omega, \mathbb{R}^{Nn})$ *and* $g \in L^2(\Omega, \mathbb{R}^N)$. *Then we have the implications:*

(i) *If* $f \in L^{2,\lambda}(\Omega, \mathbb{R}^{Nn})$, $g \in L^{2,\max\{0,\lambda-2\}}(\Omega, \mathbb{R}^N)$ *for some* $\lambda \in (0, n)$, *then we have* $Du \in L^{2,\lambda}_{\mathrm{loc}}(\Omega, \mathbb{R}^{Nn})$, *and for every subset* $\Omega' \Subset \Omega$ *there holds*

$$\|Du\|_{L^{2,\lambda}(\Omega',\mathbb{R}^{Nn})} \leq c\big(\|Du\|_{L^2(\Omega,\mathbb{R}^{Nn})} + \|f\|_{L^{2,\lambda}(\Omega,\mathbb{R}^{Nn})}$$
$$+ \|g\|_{L^{2,\max\{0,\lambda-2\}}(\Omega,\mathbb{R}^N)}\big)$$

for a constant c *depending only on* n, N, L, λ, *and* $\mathrm{dist}(\Omega', \partial\Omega)$;

(ii) *If* $f \in \mathcal{L}^{2,\lambda}(\Omega, \mathbb{R}^{Nn})$, $g \in L^{2,\max\{0,\lambda-2\}}(\Omega, \mathbb{R}^N)$ *for some* $\lambda \in (0, n+2)$, *then we have* $Du \in \mathcal{L}^{2,\lambda}_{\mathrm{loc}}(\Omega, \mathbb{R}^{Nn})$, *and for every subset* $\Omega' \Subset \Omega$ *there holds*

$$\|Du\|_{\mathcal{L}^{2,\lambda}(\Omega',\mathbb{R}^{Nn})} \leq c\big(\|Du\|_{L^2(\Omega,\mathbb{R}^{Nn})} + \|f\|_{\mathcal{L}^{2,\lambda}(\Omega,\mathbb{R}^{Nn})}$$
$$+ \|g\|_{L^{2,\max\{0,\lambda-2\}}(\Omega,\mathbb{R}^N)}\big)$$

for a constant c *depending only on* n, N, L, λ, *and* $\mathrm{dist}(\Omega', \partial\Omega)$.

Remark 4.15 Under the same assumptions as in Corollary 4.14, we emphasize that the following Hölder regularity results are immediate. In the setting (i) with $\lambda \in (n-2, n)$, we have local Hölder regularity of u with $u \in C^{0,(\lambda+2-n)/2}(\Omega, \mathbb{R}^N)$ (obtained by the local version of Corollary 1.58, see Remark 1.60). In the setting (ii) with $\lambda \in (n, n+2)$, we have local Hölder regularity of Du with $Du \in C^{0,(\lambda-n)/2}(\Omega, \mathbb{R}^{Nn})$ (as a direct consequence of the Campanato isomorphy in Theorem 1.27).

Inhomogeneous systems with continuous coefficients Finally, we deal with systems of the form (4.2) with uniformly continuous coefficients, i.e., we suppose that there exists a modulus of continuity $\omega \colon \mathbb{R}_0^+ \to \mathbb{R}_0^+$ with $\lim_{t \searrow 0} \omega(t) = \omega(0) = 0$ such that

$$|a(x) - a(y)| \leq \omega(|x - y|) \qquad \text{for all } x, y \in \Omega. \tag{4.12}$$

With the technique of *"freezing the coefficients"*, which is again a perturbation argument, we immediately find the optimal decay estimates for weak solutions to such systems.

Lemma 4.16 (Decay estimates III; Campanato) *Let $u \in W^{1,2}(\Omega, \mathbb{R}^N)$ be a weak solution to the system (4.2) with continuous coefficients $a \in C^0(\Omega, \mathbb{R}^{Nn \times Nn})$ satisfying (4.3), (4.4), and (4.12), let $f \in L^2(\Omega, \mathbb{R}^{Nn})$ and $g \in L^2(\Omega, \mathbb{R}^N)$. Then for all balls $B_r(x_0) \subset B_R(x_0) \subset \Omega$ we have*

$$\int_{B_r(x_0)} |Du|^2 \, dx \le c \Big[\Big(\Big(\frac{r}{R}\Big)^n + \omega(R)^2 \Big) \int_{B_R(x_0)} |Du|^2 \, dx$$
$$+ \int_{B_R(x_0)} \big(|f|^2 + R^2|g|^2 \big) \, dx \Big]$$

and

$$\int_{B_r(x_0)} |Du - (Du)_{B_r(x_0)}|^2 \, dx$$
$$\le c \Big[\Big(\frac{r}{R}\Big)^{n+2} \int_{B_R(x_0)} |Du - (Du)_{B_R(x_0)}|^2 \, dx + \omega(R)^2 \int_{B_R(x_0)} |Du|^2 \, dx$$
$$+ \int_{B_R(x_0)} \big(|f - (f)_{B_R(x_0)}|^2 + R^2|g|^2 \big) \, dx \Big],$$

with constants c depending only on n, N, and L.

Proof We fix $B_R(x_0) \subset \Omega$ and observe that u is a weak solution to the inhomogeneous system

$$\mathrm{div}\, \big(a(x_0)Du \big) = \mathrm{div}\, \big((a(x_0) - a(x))Du + f \big) - g$$

with constant, elliptic coefficients $a(x_0)$. Hence, the desired inequalities follow immediately from Lemma 4.13 and the uniform continuity condition (4.12). $\qquad \square$

With the iteration Lemma B.3, we thus obtain the following Morrey- and Campanato-space regularity results, which are known as Schauder estimates.

Corollary 4.17 *Let $u \in W^{1,2}(\Omega, \mathbb{R}^N)$ be a weak solution to the system (4.2) with continuous coefficients $a \in C^0(\Omega, \mathbb{R}^{Nn \times Nn})$ satisfying (4.3), (4.4),* *and* (4.12), $f \in L^2(\Omega, \mathbb{R}^{Nn})$ *and* $g \in L^2(\Omega, \mathbb{R}^N)$. *Then we have the implications:*

(i) *If* $f \in L^{2,\lambda}(\Omega, \mathbb{R}^{Nn})$ *and* $g \in L^{2,\max\{0,\lambda-2\}}(\Omega, \mathbb{R}^N)$ *for some* $\lambda \in (0, n)$, *then we have* $Du \in L^{2,\lambda}_{\mathrm{loc}}(\Omega, \mathbb{R}^{Nn})$, *and for every subset* $\Omega' \Subset \Omega$ *there holds*

$$\|Du\|_{L^{2,\lambda}(\Omega',\mathbb{R}^{Nn})} \le c\big(\|Du\|_{L^2(\Omega,\mathbb{R}^{Nn})} + \|f\|_{L^{2,\lambda}(\Omega,\mathbb{R}^{Nn})}$$
$$+ \|g\|_{L^{2,\max\{0,\lambda-2\}}(\Omega,\mathbb{R}^N)}\big)$$

   for a constant c *depending only on* n, N, L, λ, ω, *and* $\mathrm{dist}(\Omega', \partial\Omega)$;

(ii) *If* $\omega(t) \le t^{(\lambda-n)/2}$, $f \in \mathcal{L}^{2,\lambda}(\Omega, \mathbb{R}^{Nn})$ *and* $g \in L^{2,\lambda-2}(\Omega, \mathbb{R}^N)$ *for some* $\lambda \in (n, n+2)$, *then we have* $Du \in \mathcal{L}^{2,\lambda}_{\mathrm{loc}}(\Omega, \mathbb{R}^{Nn}) \simeq C^{0,(\lambda-n)/2}(\Omega, \mathbb{R}^{Nn})$, *and for every subset* $\Omega' \Subset \Omega$ *there holds*

$$\|Du\|_{\mathcal{L}^{2,\lambda}(\Omega',\mathbb{R}^{Nn})} \le c\big(\|Du\|_{L^2(\Omega,\mathbb{R}^{Nn})} + \|f\|_{\mathcal{L}^{2,\lambda}(\Omega,\mathbb{R}^{Nn})}$$
$$+ \|g\|_{L^{2,\lambda-2}(\Omega,\mathbb{R}^N)}\big)$$

   for a constant c *depending only on* n, N, L, λ, ω, *and* $\mathrm{dist}(\Omega', \partial\Omega)$.

Also here, taking into account the higher-differentiability result from Theorem 4.9 and relying on the differentiated system, we can generalize these Morrey- and Campanato-type regularity results to higher order.

Theorem 4.18 *Let* $k \in \mathbb{N}$ *and consider a weak solution* $u \in W^{1,2}(\Omega, \mathbb{R}^N)$ *to the system* (4.2) *with coefficients* $a \in C^k(\Omega, \mathbb{R}^{Nn \times Nn})$ *satisfying* (4.3), $f \in W^{k,2}(\Omega, \mathbb{R}^{Nn})$ *and* $g \in W^{k-1,2}(\Omega, \mathbb{R}^N)$. *Then we have the implications (with corresponding estimates):*

(i) *If* $D^k f \in L^{2,\lambda}(\Omega, \mathbb{R}^{Nn^{k+1}})$ *and* $D^{k-1}g \in L^{2,\lambda}(\Omega, \mathbb{R}^{Nn^{k-1}})$ *for some* $\lambda \in (0, n)$, *then we have* $D^{k+1}u \in L^{2,\lambda}_{\mathrm{loc}}(\Omega, \mathbb{R}^{Nn^{k+1}})$;

(ii) *If* $a \in C^{k,(\lambda-n)/2}(\Omega, \mathbb{R}^{Nn \times Nn})$, $D^k f \in \mathcal{L}^{2,\lambda}(\Omega, \mathbb{R}^{Nn^{k+1}})$ *and* $D^{k-1}g \in \mathcal{L}^{2,\lambda}(\Omega, \mathbb{R}^{Nn^{k-1}})$ *for some* $\lambda \in (n, n+2)$, *then we have* $D^{k+1}u \in \mathcal{L}^{2,\lambda}_{\mathrm{loc}}(\Omega, \mathbb{R}^{Nn^{k+1}}) \simeq C^{0,(\lambda-n)/2}(\Omega, \mathbb{R}^{Nn^{k+1}})$.

Remark 4.19 Due to Theorem 4.9 and the previous result, we know that both differentiability and decay properties of the inhomogeneity are carried over to the gradient of weak solutions to the associated linear system. The same actually holds for integrability properties, i.e., whenever we have $f \in L^p(\Omega, \mathbb{R}^{Nn})$, $g \in L^{np/(n+p)}(\Omega, \mathbb{R}^N)$ for some $p \in (2, \infty)$ and uniformly continuous coefficients a, then $Du \in L^p_{\mathrm{loc}}(\Omega, \mathbb{R}^{Nn})$. This result is referred to as L^p-*theory* and is for example proved in [40, Chapter 10.4].

4.3 Approaches for Partial $C^{0,\alpha}$-Regularity

We next study weak solutions to particular quasilinear systems of the form

$$\operatorname{div}\big(a(x,u)Du\big) = 0 \qquad \text{in } \Omega. \tag{4.13}$$

These systems are linear in the gradient variable and play the role of model systems illustrating some essential features (cp. the counterexamples to full regularity in Sect. 4.1) and techniques in the vectorial setting, specifically concerning the regularity properties of their weak solutions. The aim of this section is to present three different approaches to partial $C^{0,\alpha}$-continuity of such solutions, a regularity result, which was first established by Giusti and Miranda in [41]. As before, we assume ellipticity and boundedness of the coefficients in the sense of (4.6) and (4.7). Furthermore, we require some more regularity on the coefficients than being a Carathéodory function, namely we assume uniform continuity on $\Omega \times \mathbb{R}^N$. This means that there exists of a modulus of continuity $\omega \colon \mathbb{R}_0^+ \to \mathbb{R}_0^+$ (concave and monotonically non-decreasing) satisfying $\lim_{t \searrow 0} \omega(t) = \omega(0) = 0$ such that

$$|a(x,u) - a(\tilde{x},\tilde{u})| \le \omega(|x - \tilde{x}| + |u - \tilde{u}|) \tag{4.14}$$

for all $x, \tilde{x} \in \Omega$ and all $u, \tilde{u} \in \mathbb{R}^N$.

Partial regularity as optimal regularity result In view of Example 4.3 of Giusti and Miranda, we cannot expect that a given weak solution to a system of the form (4.13) is everywhere regular in Ω, under the above assumptions on the coefficients. However, the irregular solution in Example 4.3 is given by the function $u(x) = x/|x|$, which has a discontinuity in only one point and is smooth everywhere else. Such a pointwise regularity in an open subset of Ω, whose complement is negligible with respect to the $\mathcal{L}^n$-measure, is called *partial regularity*. Regularity here refers to continuity of the weak solution (or of its gradient), and in order to study regularity under the aspect of optimality, we now introduce the (open) α-regular set of a measurable function $f \colon \Omega \to \mathbb{R}^N$ via

$$\operatorname{Reg}_\alpha(f) := \big\{ x_0 \in \Omega \colon f \text{ is locally continuous}$$
$$\text{near } x_0 \text{ with Hölder exponent } \alpha \big\}$$

for $\alpha \in [0,1]$, and the singular set of f as its complement in Ω, i.e.

$$\operatorname{Sing}_\alpha(f) := \Omega \setminus \operatorname{Reg}_\alpha(f).$$

Note that we have the obvious inclusion $\operatorname{Reg}_{\alpha_1}(f) \supseteq \operatorname{Reg}_{\alpha_2}(f)$ whenever $\alpha_1 \le \alpha_2$.

Some general comments on the strategy of proof Apart from proving almost everywhere regularity of a weak solution u, almost all partial regularity results contain also a *regularity improvement* which states the equivalence $\mathrm{Reg}_0(u) = \mathrm{Reg}_\alpha(u)$ for some $\alpha > 0$ (or the corresponding equality for the regular set of the gradient Du instead of u). In this section we will give three different proofs of the partial $C^{0,\alpha}$-regularity result for weak solutions to (4.13). Since the setting is quite simple, we can get to know the essential components of these partial regularity proofs.

The first main ingredient is a *Caccioppoli inequality* which allows to control a suitable norm of the derivative of the weak solution by a norm of the solution itself, possibly on a larger set. For the setting under consideration here, such an inequality was already provided in Proposition 4.5. The second ingredient is an *excess decay estimate*, where the excess is defined as the averaged mean-square deviation from the mean of the relevant function, here for the solution itself, over a ball, that is

$$E(u; x_0, \varrho) := \fint_{B_\varrho(x_0)} |u - (u)_{B_\varrho(x_0)}|^2 \, dx \qquad (4.15)$$

with $B_\varrho(x_0) \subset \Omega$. The objective is to determine the scaling behavior of the excess with respect to the radius, in a similar form as in Lemma 4.11, where we have achieved $E(u; x_0, r) \leq c(r/R)^2 E(u; x_0, R)$ for every weak solution to a linear system with constant coefficients and all radii $r \leq R$. The importance of these excess decay estimates (either for any ratio or for a fixed ratio of radii) becomes clear in view of its relation to Campanato spaces and their equivalence to Hölder spaces, see Theorem 1.27. Obviously, due to the possible emergence of discontinuities, such excess decay estimates will only be true provided that we choose a "good" point (and hopefully one can justify that almost every point is actually a good one). Good here refers to the possibility of applying a *comparison principle* with a weak solution to a suitably linearized system – as we have already observed, these weak solutions enjoy optimal decay estimates, which can then be transferred (at least up to a certain degree) to the original solution. For the implementation of this linearization (and thus the proof of the excess decay estimate) there are several different approaches, which will be discussed in detail in the next three subsections: the blow-up technique employed by Giusti and Miranda [41], the method of $\mathcal{A}$-harmonic approximation used by Duzaar and Grotowski [21], and finally the direct approach implemented by Giaquinta and Giusti [31] and by Ivert [48]. Once these initial excess decay estimates for the weak solution are established, we can proceed to the proof of the partial $C^{0,\alpha}$-regularity result. This is stated in Theorem 4.23 along with the *characterization of singular points* (namely as the set of points for which the comparison with a linearized system might fail). In what follows, we stay close to the presentation of Duzaar's lecture series [20].

4.3.1 The Blow-Up Technique

We first explain the proof of the partial $C^{0,\alpha}$-regularity given by Giusti and Miranda in [41], which is based on an indirect approach, called the "blow-up technique". This technique traces its origins back to works of De Giorgi [14] and of Almgren [2] in the context of regularity of minimal surfaces and can roughly be described as follows. One proceeds by contradiction and therefore assumes that there exist a sequence of balls $(B_{R_j}(x_j))_{j\in\mathbb{N}}$ and a sequence of weak solutions $(u_j)_{j\in\mathbb{N}}$ for which the excess decay estimate fails. One then rescales and translates these functions in order to obtain a related sequence of functions $(v_j)_{j\in\mathbb{N}}$ defined on the unit ball (which has the interpretation of a blown-up neighbourhood of each x_j), each of which still fails to satisfy the decay estimate. This new sequence is bounded uniformly in L^2 and hence has a subsequence converging weakly in L^2. Moreover, its weak limit is a weak solution to a linear system obtained during the blow-up process. As a consequence of the relevant linear theory, this limit function now *does* satisfy the desired excess decay estimate. Since the excess is given in terms of an L^2-estimate of v, this contradicts the fact that the sequence $(v_j)_{j\in\mathbb{N}}$ fails to satisfy the decay estimate, provided that one can improve the a priori weak convergence of the sequence $(v_j)_{j\in\mathbb{N}}$ to *strong* convergence in L^2.

To start the detailed discussion of the regularity proof of Giusti and Miranda, we first provide the improvement of weak to strong convergence for a suitable sequence of weak solutions.

Lemma 4.20 ([41], Lemma 2) *Let $(b_j)_{j\in\mathbb{N}}$ be a sequence of bilinear forms such that for every $j \in \mathbb{N}$ the functions $b_j\colon B_1 \to \mathbb{R}^{Nn\times Nn}$ are measurable, bounded and elliptic in the sense of*

$$b_j(x)\xi \cdot \xi \geq |\xi|^2$$

$$b_j(x)\xi \cdot \tilde{\xi} \leq L|\xi||\tilde{\xi}|$$

for almost every $x \in B_1$, all $\xi,\tilde{\xi} \in \mathbb{R}^{Nn}$ and some $L \geq 1$. Suppose that b_j converges pointwise almost everywhere in B_1 to some bilinear form $b\colon B_1 \to \mathbb{R}^{Nn\times Nn}$. Let further $(u_j)_{j\in\mathbb{N}}$ be a sequence in $W^{1,2}(B_1,\mathbb{R}^N)$ such that u_j solves the system $\mathrm{div}\,(b_j(x)Du_j) = 0$ in B_1 in the weak sense for every $j \in \mathbb{N}$, and which converges weakly in $L^2(B_1,\mathbb{R}^N)$ to a function $u \in L^2(B_1,\mathbb{R}^N)$. Then $u \in W^{1,2}_{\mathrm{loc}}(B_1,\mathbb{R}^N)$, and we have

(i) *$u_j \to u$ strongly in $L^2(B_\varrho,\mathbb{R}^N)$, $Du_j \rightharpoonup Du$ weakly in $L^2(B_\varrho,\mathbb{R}^{Nn})$ for every $\varrho < 1$;*

(ii) *u solves the system $\mathrm{div}\,(b(x)Du) = 0$ in B_1 in the weak sense.*

Proof The sequence $(u_j)_{j\in\mathbb{N}}$ converges weakly in $L^2(B_1,\mathbb{R}^N)$. Hence, in particular, it is bounded in $L^2(B_1,\mathbb{R}^N)$. By the Caccioppoli inequality from

Proposition 4.5, the sequence $(u_j)_{j\in\mathbb{N}}$ is then bounded in $W^{1,2}(B_\varrho, \mathbb{R}^N)$ for every $\varrho < 1$, since we have

$$\int_{B_\varrho} |Du_j|^2\, dx \le c(L)(1-\varrho)^{-2} \int_{B_1} |u_j|^2\, dx \le c(L,\varrho)\sup_{j\in\mathbb{N}} \|u_j\|^2_{L^2(B_1,\mathbb{R}^N)}$$

with a constant c that is independent of $j \in \mathbb{N}$. By weak precompactness of $W^{1,2}(B_\varrho, \mathbb{R}^N)$ and Rellich's Theorem 1.54, we can therefore extract a subsequence of $(u_j)_{j\in\mathbb{N}}$ that converges weakly in $W^{1,2}(B_\varrho, \mathbb{R}^N)$ and strongly in $L^2(B_\varrho, \mathbb{R}^N)$. Due to the a priori weak convergence $u_j \rightharpoonup u$ in $L^2(B_1, \mathbb{R}^N)$, the whole sequence actually converges, and the limit is given by u. It only remains to prove (ii). To this end we take a test function $\varphi \in C^1_0(B_1, \mathbb{R}^N)$ with $\operatorname{spt}\varphi \subset B_\varrho$ for some $\varrho < 1$. Now we write

$$\int_{B_1} bDu \cdot D\varphi\, dx = \int_{B_1} b(Du - Du_j)\cdot D\varphi\, dx$$
$$+ \int_{B_1} (b - b_j)Du_j \cdot D\varphi\, dx + \int_{B_1} b_j Du_j \cdot D\varphi\, dx$$

for $j \in \mathbb{N}$. In view of the weak convergence $Du_j \rightharpoonup Du$ in $L^2(B_\varrho, \mathbb{R}^{Nn})$, the first integral vanishes in the limit $j \to \infty$. For the second integral we compute via Hölder's inequality

$$\left| \int_{B_1} (b - b_j)Du_j \cdot D\varphi\, dx \right| \le \sup_{B_\varrho} |D\varphi| \left(\int_{B_\varrho} |b - b_j|^2\, dx \right)^{\frac{1}{2}} \left(\int_{B_\varrho} |Du_j|^2\, dx \right)^{\frac{1}{2}}$$

and observe that the right-hand side converges to zero by the pointwise convergence $b_j \to b$, the dominated convergence Theorem 1.11 and the boundedness of $(u_j)_{j\in\mathbb{N}}$ in $W^{1,2}(B_\varrho, \mathbb{R}^N)$. Finally, the third integral vanishes for every $j \in \mathbb{N}$ by assumption, since the functions u_j are weak solutions of the system $\operatorname{div}(b_j(x)Du_j) = 0$ in B_1. Hence, we obtain $\operatorname{div}(b(x)Du) = 0$ in B_1 in the weak sense as claimed in (ii), and the proof of the lemma is complete.   □

Lemma 4.21 (Excess decay estimate via blow up; [41], Lemma 4)
For every $\tau \in (0,1)$ there exist two positive constants ε_0, R_0 depending only on n, N, L, ω, and τ such that the following statement is true: if $u \in W^{1,2}(\Omega, \mathbb{R}^N)$ is a weak solution to the system (4.13) with continuous coefficients $a\colon \Omega \times \mathbb{R}^N \to \mathbb{R}^{Nn \times Nn}$ satisfying (4.6), (4.7) and (4.14), and if for some ball $B_R(x_0) \subset \Omega$ with $R \le R_0$ there holds

$$E(u; x_0, R) < \varepsilon_0^2, \qquad\qquad (4.16)$$

then we have the excess decay estimate

$$E(u; x_0, \tau R) \leq c_*(n, N, L)\tau^2 E(u; x_0, R). \tag{4.17}$$

Proof If the conclusion of the lemma were false, we could find a number $\tau \in (0,1)$, a sequence of balls $(B_{R_j}(x_j))_{j \in \mathbb{N}}$ contained in Ω with $R_j \searrow 0$, a sequence of uniformly continuous bilinear forms $(a_j)_{j \in \mathbb{N}}$ with $a_j \colon \Omega \times \mathbb{R}^N \to \mathbb{R}^{Nn \times Nn}$ satisfying (4.6), (4.7) and (4.14) for all $j \in \mathbb{N}$, and a sequence of functions $(u_j)_{j \in \mathbb{N}}$ in $W^{1,2}(\Omega, \mathbb{R}^N)$ with the following properties:

(i) u_j is weak solution to the system $\operatorname{div}(a_j(x, u_j)Du_j) = 0$ in Ω,
(ii) $E(u_j; x_j, R_j) = \varepsilon_j^2 \searrow 0$,
(iii) $E(u_j; x_j, \tau R_j) > c_*(n, N, L)\tau^2 \varepsilon_j^2$,

for all $j \in \mathbb{N}$ and for some constant $c_*(n, N, L)$ that will be fixed at the end of the proof. Now we rescale and normalize the functions u_j and accordingly the coefficients a_j in order to obtain a sequence $(v_j)_{j \in \mathbb{N}}$ of functions in $W^{1,2}(B_1, \mathbb{R}^N)$ (having in addition average zero in B_1), via

$$v_j(y) := \varepsilon_j^{-1}\big[u_j(x_j + R_j y) - (u_j)_{B_{R_j}(x_j)}\big] \qquad \text{for } y \in B_1,$$

and a sequence of bilinear forms $(b_j)_{j \in \mathbb{N}}$ in $L^\infty(B_1, \mathbb{R}^{Nn \times Nn})$, via

$$b_j(y) := a_j(x_j + R_j y, \varepsilon_j v_j(y) + (u_j)_{B_{R_j}(x_j)})$$

$$= a_j(x_j + R_j y, u_j(x_j + R_j y)) \qquad \text{for } y \in B_1,$$

for all $j \in \mathbb{N}$. Using the change of variable formula, we easily check that the previous properties of the sequence $(u_j)_{j \in \mathbb{N}}$ imply the following properties for the sequence $(v_j)_{j \in \mathbb{N}}$:

(o′) the mean value of v_j in B_1 vanishes,
(i′) v_j is a weak solution to $\operatorname{div}(b_j(y)Dv_j) = 0$ in B_1 since for every function $\psi \in C_0^1(B_1, \mathbb{R}^N)$ we have, by definition of v_j and b_j,

$$\int_{B_1} b_j(y)Dv_j(y) \cdot D\psi(y)\, dy$$

$$= \varepsilon_j^{-1} R_j \int_{B_1} a_j(x_j + R_j y, u_j(x_j + R_j y))Du_j(x_j + R_j y) \cdot D\psi(y)\, dy$$

$$= \varepsilon_j^{-1} R_j^{1-n} \int_{B_{R_j}(x_j)} a_j(x, u_j(x))Du_j(x) \cdot D\psi(R_j^{-1}(x - x_j))\, dx$$

$$= \varepsilon_j^{-1} R_j^{2-n} \int_{B_{R_j}(x_j)} a_j(x, u_j(x))Du_j(x) \cdot D\varphi(x)\, dx = 0$$

as a consequence of the weak formulation of (i), tested with the function $\varphi(x) := \psi(R_j^{-1}(x - x_j)) \in C_0^1(B_{R_j}(x_j), \mathbb{R}^N)$,

(ii') The excess of v_j in B_1 is normalized to 1 since

$$E(v_j; 0, 1) = \fint_{B_1} |v_j - (v_j)_{B_1}|^2 \, dx$$

$$= \fint_{B_1} |v_j|^2 \, dx = \varepsilon_j^{-2} E(u_j; x_j, R_j) = 1,$$

(iii') $E(v_j; 0, \tau) = \varepsilon_j^{-2} E(u_j; x_j, \tau R_j) > c_* \tau^2$,

for all $j \in \mathbb{N}$. In view of (ii') and the boundedness (4.7) of the bilinear forms, we can choose a subsequence such that

$$\begin{aligned}
v_{j(\ell)} &\rightharpoonup v && \text{weakly in } L^2(B_1, \mathbb{R}^N), \\
\varepsilon_{j(\ell)} v_{j(\ell)} &\to 0 && \text{almost everywhere in } B_1, \\
a_{j(\ell)}\big(x_{j(\ell)}, (u_{j(\ell)})_{B_{R_{j(\ell)}}(x_{j(\ell)})}\big) &\to b && \text{in } \mathbb{R}^{Nn \times Nn}.
\end{aligned}$$

Moreover, for the limit function v there hold $(v)_{B_1} = 0$ and, by weak semicontinuity of the norm,

$$\int_{B_1} |v|^2 \, dx \le 1. \tag{4.18}$$

We now observe that all bilinear forms of the sequence $(b_j)_{j\in\mathbb{N}}$ are measurable, and they further satisfy (4.6) and (4.7). Moreover, due to the uniform continuity assumption on the coefficients a_j, we find the estimate

$$|b_j(y) - b| \le |a_j(x_j + R_j y, \varepsilon_j v_j(y) + (u_j)_{B_{R_j}(x_j)}) - a_j(x_j, (u_j)_{B_{R_j}(x_j)})|$$

$$+ |a_j(x_j, (u_j)_{B_{R_j}(x_j)}) - b|$$

$$\le \omega(R_j + \varepsilon_j |v_j(y)|) + |a_j(x_j, (u_j)_{B_{R_j}(x_j)}) - b|. \tag{4.19}$$

Both terms on the right-hand side vanish in the limit for the subsequence $j(\ell)$, for almost every $y \in B_1$. This means that the sequence $b_{j(\ell)}$ converges almost everywhere to b, and in combination with the above properties, we have therefore verified all assumptions of the previous Lemma 4.20. Consequently, we obtain in particular strong convergence $v_{j(\ell)} \to v$ in $L^2(B_\tau, \mathbb{R}^N)$ for the number $\tau < 1$ fixed at the beginning of the proof, and v is a weak solution to the homogeneous linear system $\mathrm{div}\,(bDv) = 0$ in B_1, where the coefficients b are *constant* by construction. Thus, Lemma 4.11 from the linear theory provides the decay estimate

$$E(v;0,\tau) = \fint_{B_\tau} |v - (v)_{B_\tau}|^2 \, dx \leq c(n,N,L)\tau^2 \fint_{B_1} |v|^2 \, dx \leq c(n,N,L)\tau^2 \, .$$

In view of the strong convergence $v_{j(\ell)} \to v$ in $L^2(B_\tau, \mathbb{R}^N)$, we find $\ell_0 \in \mathbb{N}$ such that

$$E(v_{j(\ell)};0,\tau) \leq 2c(n,N,L)\tau^2 \qquad \text{for all } \ell \geq \ell_0 \, .$$

For the choice $c_* := 2c(n,N,L)$ of the constant in property (iii) at the beginning of the proof, this is a contradiction to (iii′), and hence, the lemma is proved.   □

Remark 4.22 In fact, an excess decay estimate of the form (4.17) is also obtained under the assumption of continuous, but not necessarily uniformly continuous coefficients, i.e., if we assume

$$|a(x,u) - a(\tilde{x},\tilde{u})| \leq \tilde{\omega}(|u| + |\tilde{u}|, |x - \tilde{x}| + |u - \tilde{u}|) \qquad (4.20)$$

for all $x, \tilde{x} \in \Omega$, all $u, \tilde{u} \in \mathbb{R}^N$, and where $\tilde{\omega} \colon \mathbb{R}_0^+ \times \mathbb{R}_0^+ \to \mathbb{R}_0^+$ is a function with the following properties: $t \mapsto \tilde{\omega}(M,t)$ is a modulus of continuity with $\lim_{t \searrow 0} \tilde{\omega}(M,t) = \tilde{\omega}(M,0) = 0$ for every $M \in \mathbb{R}_0^+$, and $M \to \tilde{\omega}(M,t)$ is non-decreasing for every $t \in \mathbb{R}_0^+$. In this case, we additionally need to assume a bound of the form $|(u)_{B_R(x_0)}| \leq M$ on the mean values (consequently, the numbers ε_0 and R_0 will also depend on this number M). This assumption still allows us to deduce the pointwise convergence $b_j \to b$ almost everywhere in B_1 as in (4.19), and then the statement follows exactly as in the situation of uniformly continuous coefficients.

The previous excess decay estimate constitutes the main component of the proof of the partial $C^{0,\alpha}$-regularity result for weak solutions, which was first achieved by Giusti and Miranda in [41] and by Morrey in [66]. In fact, once such excess decay estimates are proven, the partial regularity result follows essentially from an iteration argument, which is given right away (and which is in fact the same for the three presented comparison approaches leading to the excess decay estimate in Lemma 4.21).

Theorem 4.23 (Giusti and Miranda, Morrey)   *Let $u \in W^{1,2}(\Omega, \mathbb{R}^N)$ be a weak solution to the system* (4.13) *with continuous coefficients $a \colon \Omega \times \mathbb{R}^N \to \mathbb{R}^{Nn \times Nn}$ satisfying* (4.6), (4.7) *and* (4.14). *Then we have the characterization of the singular set via*

$$\mathrm{Sing}_0(u) = \left\{ x_0 \in \Omega \colon \liminf_{\varrho \searrow 0} \fint_{\Omega(x_0,\varrho)} |u - (u)_{\Omega(x_0,\varrho)}|^2 \, dx > 0 \right\}$$

and in particular $\mathcal{L}^n(\mathrm{Sing}_0(u)) = 0$. Moreover, for every $\alpha \in (0,1)$ there holds $\mathrm{Reg}_0(u) = \mathrm{Reg}_\alpha(u)$, i.e. $u \in C^{0,\alpha}(\mathrm{Reg}_0(u), \mathbb{R}^N)$.

Proof We fix $\alpha \in (0,1)$ and choose $\tau = \tau(n, N, L, \alpha) \in (0,1)$ such that $c_*(n, N, L)\tau^{2(1-\alpha)} \leq 1$, where c_* is the constant determined in Lemma 4.21. We now need to prove that if $x_0 \in \Omega$ is a point for which

$$\liminf_{\varrho \searrow 0} \fint_{\Omega(x_0,\varrho)} |u - (u)_{\Omega(x_0,\varrho)}|^2 \, dx = 0$$

is satisfied, then we have $x_0 \in \mathrm{Reg}_0(u)$ (while the reverse implication is obviously true) and, moreover, also $x_0 \in \mathrm{Reg}_\alpha(u)$ holds. We first observe that, by assumption, we can find a ball $B_R(x_0) \Subset \Omega$ centered in x_0 and with $R \leq R_0$ such that

$$E(u; x_0, R) < \varepsilon_0^2$$

holds, where $\varepsilon_0 = \varepsilon_0(n, N, L, \omega, \alpha)$ is the smallness constant from Lemma 4.21. Consequently, the excess decay estimate (4.17) and the choice of τ imply

$$E(u; x_0, \tau R) \leq c_* \tau^2 E(u; x_0, R) \leq \tau^{2\alpha} E(u; x_0, R) < \varepsilon_0^2.$$

Therefore, the smallness condition (4.17) is also satisfied on the smaller ball $B_{\tau R}(x_0)$, and then, by induction, on each ball $B_{\tau^k R}(x_0)$ for $k \in \mathbb{N}_0$, with the estimate

$$E(u; x_0, \tau^k R) \leq \tau^{2\alpha k} E(u; x_0, R).$$

For an arbitrary radius $r \in (0, R)$ we find the final decay estimate by interpolation, which means that we determine first a number $k \in \mathbb{N}_0$ such that $\tau^{k+1} R < r \leq \tau^k R$ holds, and we then compute

$$E(u; x_0, r) \leq \left(\frac{\tau^k R}{r}\right)^n E(u; x_0, \tau^k R)$$

$$\leq \left(\frac{r}{R}\right)^{2\alpha} \tau^{-n-2\alpha(k+1)+2\alpha k} E(u; x_0, R)$$

$$\leq c(n, N, L, \alpha) \left(\frac{r}{R}\right)^{2\alpha} E(u; x_0, R).$$

In order to show a local Campanato estimate in $\mathcal{L}^{2,n+2\alpha}$ (and hence local α-Hölder continuity) we need to verify such an estimate not only for the point x_0, but for all points in a neighbourhood of x_0. For this purpose, we note, by the absolute continuity of the integral, that the function $x \mapsto E(u; x, R)$ is continuous for every fixed radius $R > 0$. Therefore, there exists a ball $B_\delta(x_0)$ such that both the inclusion $B_R(y) \Subset \Omega$ and the estimate $E(u; y, R) < \varepsilon_0^2$ are satisfied for all points $y \in B_\delta(x_0)$. In this way, we obtain

$$E(u; y, r) \leq c(n, N, L, \omega, \alpha)\left(\frac{r}{R}\right)^{2\alpha} \qquad \text{for all } y \in B_\delta(x_0) \text{ and } r \in (0, R),$$

i.e., we end up with $u \in \mathcal{L}^{2, n+2\alpha}(B_\delta(x_0), \mathbb{R}^N) \simeq C^{0,\alpha}(\overline{B_\delta(x_0)}; \mathbb{R}^N)$ (see Theorem 1.27). Finally, we observe that $\mathrm{Sing}_0(u)$ is of $\mathcal{L}^n$-measure zero, as a direct consequence of the characterization of the singular points and Lebesgue's differentiation Theorem (see Corollary 1.13).   □

Remark 4.24 If the coefficients are more regular, then we can also deduce higher regularity of weak solutions $u \in W^{1,2}(\Omega, \mathbb{R}^N)$ to the system (4.13), under the assumptions of Theorem 4.23. More precisely, we have:

(i) If the coefficients $a \colon \Omega \times \mathbb{R}^N \to \mathbb{R}^{Nn \times Nn}$ are uniformly Hölder continuous, i.e., if they satisfy (4.14) with a Hölder continuous modulus of continuity $\omega(t) \leq \min\{1, t^\alpha\}$ for all $t \in \mathbb{R}_0^+$ and some $\alpha \in (0, 1)$, then there holds $\mathrm{Reg}_0(u) = \mathrm{Reg}_\alpha(Du)$. This is a consequence of the linear theory, see Corollary 4.17, applied with coefficients $\tilde{a}(x) := a(x, u(x))$ on the regular set of u.

(ii) If the coefficients $a \colon \Omega \times \mathbb{R}^N \to \mathbb{R}^{Nn \times Nn}$ are k-times differentiable with respect to both variables and if the k-th order derivatives are uniformly Hölder continuous with exponent $\alpha \in (0, 1)$, then we obtain $\mathrm{Reg}_0(u) = \mathrm{Reg}_\alpha(D^{k+1}u)$ with similar arguments.

With the characterization of the singular points in Theorem 4.23 and the estimate for the Hausdorff dimension of the set of non-Lebesgue points of a Sobolev function from Proposition 1.76, we further get an improved estimate on the size of the singular set.

Corollary 4.25 *Let $u \in W^{1,2}(\Omega, \mathbb{R}^N)$ be a weak solution to the system* (4.13) *under the assumptions of Theorem 4.23. Then we have* $\dim_{\mathcal{H}}(\mathrm{Sing}_0(u)) \leq n - 2$.

4.3.2 The Method of A-Harmonic Approximation

We now address the first alternative approach for proving partial regularity for weak solutions to the system (4.13). More precisely, we now wish to obtain the excess decay estimate, which was established in the proof of Lemma 4.21 before via the blow-up technique, by the method of $\mathcal{A}$-harmonic approximation. This is again an implementation of a comparison principle, which allows us to transfer decay properties from a suitable solution of a linearized system (for which we have good a priori estimates, due to the linear theory from Sect. 4.2) to the weak solutions of the original system.

In the setting of partial regularity theory for elliptic systems this approach was first implemented by Duzaar and Grotowski [21] (but also employed by Duzaar and Steffen [23] for proving an interior ε-regularity theorem for rectifiable currents in the context of geometric measure theory). Since then it has been applied in various situations concerning partial regularity of solutions to elliptic and parabolic problems. To explain the idea of $\mathcal{A}$-harmonic approximation we first define $\mathcal{A}$-harmonic functions (extending the definition of harmonic functions):

Definition 4.26 Let $\mathcal{A} \in \mathbb{R}^{Nn \times Nn}$. A function $h \in W^{1,1}(\Omega, \mathbb{R}^N)$ is called $\mathcal{A}$-harmonic if it satisfies

$$\int_\Omega \mathcal{A}Dh \cdot D\varphi \, dx = 0 \qquad \text{for all } \varphi \in C_0^1(\Omega, \mathbb{R}^N).$$

We now explain the main ingredient of the $\mathcal{A}$-harmonic approximation technique, namely the $\mathcal{A}$-harmonic approximation (see Lemma 4.27 below), and we describe how it will be employed in our setting. The $\mathcal{A}$-harmonic approximation is inspired by Simon's proof of the regularity theorem of Allard, cf. [76, Section 23], and extends the method of harmonic approximation (that is, an approximation with solutions to the Laplace equation) of De Giorgi [14] in a natural way to bounded elliptic operators with constant coefficients. The central idea (and statement of this $\mathcal{A}$-harmonic approximation lemma) is that if a function u is in some quantified way "close" to being $\mathcal{A}$-harmonic, in the sense that the integral in Definition 4.26 is small in terms of $\sup |D\varphi|$ for all functions $\varphi \in C_0^1(\Omega, \mathbb{R}^N)$, then u is actually close to an $\mathcal{A}$-harmonic function h in the L^2-sense. Since h, as a solution to an elliptic system with constant coefficients, is smooth and satisfies optimal decay estimates, this L^2-closeness is sufficient to establish the desired excess decay estimates for u. With this argument in mind, one needs to find a criterion that guarantees that a weak solution to the elliptic system (4.13) is close to being $\mathcal{A}$-harmonic for a suitable $\mathcal{A} \in \mathbb{R}^{Nn \times Nn}$. This will be achieved in the second Lemma 4.28, with $\mathcal{A}$ being a linearization of the original (possibly nonlinear) system and under the hypothesis that we are in a suitable neighbourhood of a "good" point (with small initial excess).

We now give the precise statement of the announced $\mathcal{A}$-harmonic approximation lemma. The proof that we present here follows [21, Lemma 2.1] (see also [22, Lemma 2.1]).

Lemma 4.27 (De Giorgi; Duzaar and Grotowski) *Let $L \geq 1$ be a fixed constant, $n, N \in \mathbb{N}$ with $n \geq 2$ and $B_\varrho(x_0) \subset \mathbb{R}^n$. For every $\varepsilon > 0$ there exists $\delta = \delta(n, N, L, \varepsilon) > 0$ with the following property: if $\mathcal{A}$ is a constant bilinear form on $\mathbb{R}^{Nn}$ which is elliptic with (4.3) and bounded by L with (4.4), and if $u \in W^{1,2}(B_\varrho(x_0), \mathbb{R}^N)$ satisfies*

$$\varrho^{2\gamma-n} \int_{B_\varrho(x_0)} |Du|^2 \, dx \leq 1$$

(for some $\gamma \in \mathbb{R}$) and is approximately $\mathcal{A}$-harmonic in the sense of

$$\left| \varrho^{\gamma-n} \int_{B_\varrho(x_0)} \mathcal{A}Du \cdot D\varphi \, dx \right| \leq \delta \sup_{B_\varrho(x_0)} |D\varphi| \quad \text{for all } \varphi \in C_0^1(B_\varrho(x_0), \mathbb{R}^N),$$

then there exists an $\mathcal{A}$-harmonic function $h \in W^{1,2}(B_\varrho(x_0), \mathbb{R}^N)$ which satisfies

$$\varrho^{2\gamma-n-2} \int_{B_\varrho(x_0)} |u - h|^2 \, dx \leq \varepsilon \quad \text{and} \quad \varrho^{2\gamma-n} \int_{B_\varrho(x_0)} |Dh|^2 \, dx \leq 1. \quad (4.21)$$

Proof Without loss of generality we may assume $x_0 = 0$ and $\varrho = 1$. The general case follows by a rescaling argument (given at the end of the proof). If the conclusion of the lemma were false, we could find $\varepsilon > 0$, a sequence $(\mathcal{A}_j)_{j\in\mathbb{N}}$ of elliptic, bounded bilinear forms on $\mathbb{R}^{Nn}$ which all satisfy (4.3) and (4.4), a sequence of functions $(u_j)_{j\in\mathbb{N}}$ in $W^{1,2}(B_1, \mathbb{R}^N)$ such that for every $j \in \mathbb{N}$ there hold

$$\int_{B_1} |Du_j|^2 \, dx \leq 1$$

and

$$\left| \int_{B_1} \mathcal{A}_j Du_j \cdot D\varphi \, dx \right| \leq k^{-1} \sup_{B_1} |D\varphi| \quad \text{for all } \varphi \in C_0^1(B_1, \mathbb{R}^N), \quad (4.22)$$

but the assertion of the lemma fails, i.e., the inequality

$$\int_{B_1} |u_j - h_j|^2 \, dx > \varepsilon \quad (4.23)$$

is satisfied for every $\mathcal{A}_j$-harmonic functions h_j with $\int_{B_1} |Dh_j|^2 \, dx \leq 1$ (note that this class contains all constant functions and is thus non-empty). Without loss of generality, we can further assume $(u_j)_{B_1} = 0$ (otherwise we replace u_j by $u_j - (u_j)_{B_1}$).

In view of Poincaré's inequality, the sequence $(u_j)_{j\in\mathbb{N}}$ is bounded in $W^{1,2}(B_1, \mathbb{R}^N)$. Hence, via weak compactness of $W^{1,2}(B_1, \mathbb{R}^N)$ and Rellich's Theorem 1.54, we can pass to a subsequence (again labeled by j) with the following properties:

$$\begin{aligned}
u_j &\rightharpoonup v \quad \text{weakly in } W^{1,2}(B_1, \mathbb{R}^N), \\
u_j &\to v \quad \text{strongly in } L^2(B_1, \mathbb{R}^N), \\
\mathcal{A}_j &\to \mathcal{A} \quad \text{in } \mathbb{R}^{Nn \times Nn},
\end{aligned}$$

for some function $v \in W^{1,2}(B_1, \mathbb{R}^N)$ and for a bilinear form $\mathcal{A}$ on $\mathbb{R}^{Nn}$. Several properties of v are inherited from the sequence $(u_j)_{j \in \mathbb{N}}$. First, using the lower semicontinuity of $w \mapsto \int_B |Dw|^2\, dx$ with respect to weak convergence in $W^{1,2}(B_1, \mathbb{R}^N)$, also v satisfies $\int_{B_1} |Dv|^2\, dx \leq 1$. Next, the strong convergence in $L^2(B_1, \mathbb{R}^N)$ guarantees in particular $(v)_{B_1} = 0$. Lastly, v is $\mathcal{A}$-harmonic in B_1; indeed, for an arbitrary test function $\varphi \in C_0^\infty(B_1, \mathbb{R}^N)$ we can write

$$\int_{B_1} \mathcal{A}Dv \cdot D\varphi\, dx = \int_{B_1} \mathcal{A}(Dv - Du_j) \cdot D\varphi\, dx$$
$$+ \int_{B_1} (\mathcal{A} - \mathcal{A}_j)Du_j \cdot D\varphi\, dx + \int_{B_1} \mathcal{A}_j Du_j \cdot D\varphi\, dx.$$

Now, the first term on the right-hand side vanishes in the limit $j \to \infty$ by the weak convergence of $u_j \rightharpoonup v$ in $W^{1,2}(B_1, \mathbb{R}^N)$, the second term due to the convergence of $\mathcal{A}_j \to \mathcal{A}$ and the boundedness of $(u_j)_{j \in \mathbb{N}}$ in $W^{1,2}(B_1, \mathbb{R}^N)$, and the third term by the $1/k$-approximate $\mathcal{A}_j$-harmonicity of u_j from (4.22).

In order to get a contradiction to (4.23), we now consider the Dirichlet problem $\mathcal{D}_j$ given by

$$\begin{cases} \operatorname{div}\left(\mathcal{A}_j Dv_j\right) = 0 & \text{in } B_1 \\ \qquad\qquad v_j = v & \text{on } \partial B_1 \end{cases}$$

for every $j \in \mathbb{N}$. By the Lax–Milgram Theorem A.11 (see also Remark 4.12) there exists a unique weak solution $v_j \in v + W_0^{1,2}(B_1, \mathbb{R}^N)$ to each Dirichlet problem $\mathcal{D}_j$. Using the ellipticity condition (4.3) for each bilinear form $\mathcal{A}_j$, the $\mathcal{A}_j$-harmonicity of v_j and the $\mathcal{A}$-harmonicity of v (with the test function $v_j - v$, which is admissible after an approximation argument), we see via Hölder's inequality

$$\int_{B_1} |Dv_j - Dv|^2\, dx \leq \int_{B_1} \mathcal{A}_j(Dv_j - Dv) \cdot (Dv_j - Dv)\, dx$$
$$= -\int_{B_1} \mathcal{A}_j Dv \cdot (Dv_j - Dv)\, dx$$
$$= \int_{B_1} (\mathcal{A} - \mathcal{A}_j)Dv \cdot (Dv_j - Dv)\, dx$$
$$\leq |\mathcal{A} - \mathcal{A}_j|\left(\int_{B_1} |Dv|^2\, dx\right)^{\frac{1}{2}}\left(\int_{B_1} |Dv_j - Dv|^2\, dx\right)^{1/2}.$$

In view of $\int_{B_1} |Dv|^2\, dx \leq 1$ and the convergence $\mathcal{A}_j \to \mathcal{A}$, we obtain strong convergence $Dv_k \to Dv$ in $L^2(B_1, \mathbb{R}^{Nn})$. In turn, via $v_j = v$ on ∂B_1 and

Poincaré's inequality, this can be improved to strong convergence $v_j \to v$ in $W^{1,2}(B_1, \mathbb{R}^N)$. Therefore, having the strong convergences of $v_j \to v$ as well as of $u_j \to v$ in $L^2(\Omega, \mathbb{R}^N)$ at hand, we have $v_j - u_j \to 0$ in $L^2(B_1, \mathbb{R}^N)$ and the function v_j would provide a contradiction to (4.23) if we also knew $\int_{B_1} |Dv_j|^2 \, dx \leq 1$. Since in general this bound cannot be guaranteed (it is known to be true only in the limit $j \to \infty$), we rescale the functions v_j via

$$h_j := \frac{v_j}{m_j} \quad \text{with } m_j := \max \left\{ 1, \left(\int_{B_1} |Dv_j|^2 \, dx \right)^{1/2} \right\}$$

and notice $m_j \to 1$ as $j \to \infty$. The functions h_j are still $\mathcal{A}_j$-harmonic, but now they additionally satisfy $\int_{B_1} |Dh_j|^2 \, dx \leq 1$ for every $j \in \mathbb{N}$ by construction. Moreover, we have

$$\|h_j - u_j\|_{L^2(B_1, \mathbb{R}^N)}$$
$$\leq \|h_j - v_j\|_{L^2(B_1, \mathbb{R}^N)} + \|v_j - v\|_{L^2(B_1, \mathbb{R}^N)} + \|v - u_j\|_{L^2(B_1, \mathbb{R}^N)} .$$

The first term on the right-hand side is bounded by $(1 - m_j^{-1}) \|v_j\|_{L^2(B_1, \mathbb{R}^N)}$, and due to the strong convergence $v_j \to v$ in $L^2(B_1, \mathbb{R}^N)$, the norms $\|v_j\|_{L^2(B_1, \mathbb{R}^N)}$ are bounded uniformly for all $j \in \mathbb{N}$. Therefore, the first term vanishes in the limit because of $m_j \to 1$ as $j \to \infty$. Moreover, the second and the third term vanish in the limit, due to the strong convergences $v_j \to v$ and $u_j \to v$, respectively, in $L^2(B_1, \mathbb{R}^N)$. Consequently, we infer $\|h_j - u_j\|_{L^2(B_1, \mathbb{R}^N)} \to 0$ as $j \to \infty$, which contradicts (4.23) and finishes the proof of the lemma for $\varrho = 1$ and $x_0 = 0$.

It remains to explain the rescaling argument to prove the result for an arbitrary balls $B_\varrho(x_0)$. If a function $u \in W^{1,2}(B_\varrho(x_0), \mathbb{R}^N)$ is given as in the statement, then we can define a function $U \in W^{1,2}(B_1, \mathbb{R}^N)$ via $U(y) := \varrho^{\gamma - 1} u(x_0 + \varrho y)$ for all $y \in B_1$. It is easy to verify that U satisfies the assumptions of the $\mathcal{A}$-harmonic approximation lemma on B_1. Thus, there exists an $\mathcal{A}$-harmonic function $H \in W^{1,2}(B_1, \mathbb{R}^N)$, which satisfies (4.21) (with u and h replaced by U and H). Then the $\mathcal{A}$-harmonic function $h \in W^{1,2}(B_\varrho(x_0), \mathbb{R}^N)$ defined via $h(x) := \varrho^{1-\gamma} H((x - x_0)/\varrho)$ for $x \in B_\varrho(x_0)$ provides the desired conclusion. $\qquad\square$

We now return to the setting of the previous Sect. 4.3.1 and study weak solutions $u \in W^{1,2}(\Omega, \mathbb{R}^N)$ to quasilinear systems of the form (4.13), under the permanent assumptions of ellipticity, boundedness and uniform continuity on the coefficients $a \colon \Omega \times \mathbb{R}^N \to \mathbb{R}^{Nn \times Nn}$ in the sense of (4.6), (4.7) and (4.14). These particular systems will now serve as a toy case in order to illustrate the basic ingredients of the partial regularity proofs by means of the method of $\mathcal{A}$-harmonic approximation (and we will treat the regularity of weak solutions to more general systems and of minimizers of convex variational functionals later in Sect. 5.2 with the same approach).

As explained above, we next need to establish a criterion that ensures that the weak solution is close to being $\mathcal{A}$-harmonic for a suitable bilinear form $\mathcal{A}$. This is achieved by an appropriate freezing of the original coefficients, provided that the initial excess is small, and this will allow us in the next step to give an alternative proof of the excess decay estimate in the key Lemma 4.21 (and hence of Theorem 4.23).

Lemma 4.28 (Approximate $\mathcal{A}$-harmonicity I) *Let $u \in W^{1,2}(\Omega, \mathbb{R}^N)$ be a weak solution to the system (4.13) with continuous coefficients $a \colon \Omega \times \mathbb{R}^N \to \mathbb{R}^{Nn \times Nn}$ satisfying (4.7) and (4.14). Then, for every $B_\varrho(x_0) \subset \Omega$ and all $u_0 \in \mathbb{R}^N$, we have*

$$\left| \varrho^{1-n} \int_{B_\varrho(x_0)} a(x_0, u_0) Du \cdot D\varphi \, dx \right|$$

$$\leq c(n, L) \omega^{1/2} \left(\varrho + \left(\fint_{B_\varrho(x_0)} |u - u_0|^2 \, dx \right)^{\frac{1}{2}} \right)$$

$$\times \left(\varrho^{2-n} \int_{B_\varrho(x_0)} |Du|^2 \, dx \right)^{\frac{1}{2}} \sup_{B_\varrho(x_0)} |D\varphi|$$

for all $\varphi \in C_0^1(B_\varrho(x_0), \mathbb{R}^N)$.

Proof We consider an arbitrary function $\varphi \in C_0^1(B_\varrho(x_0), \mathbb{R}^N)$ satisfying $\sup_{B_\varrho(x_0)} |D\varphi| \leq 1$ (the general result then follows after rescaling). Since u is a weak solution to the system (4.13), we first observe

$$\int_{B_\varrho(x_0)} a(x_0, u_0) Du \cdot D\varphi \, dx = \int_{B_\varrho(x_0)} [a(x_0, u_0) - a(x, u)] Du \cdot D\varphi \, dx.$$

Via the uniform continuity condition (4.14) and the boundedness condition (4.7), Hölder's and Jensen's inequality (which is applicable since ω is concave), we then derive the desired statement as follows:

$$\left| \fint_{B_\varrho(x_0)} a(x_0, u_0) Du \cdot D\varphi \, dx \right|$$

$$\leq c(L) \fint_{B_\varrho(x_0)} \omega^{1/2} \big(|x - x_0| + |u - u_0| \big) |Du| |D\varphi| \, dx$$

$$\leq c(L) \left(\fint_{B_\varrho(x_0)} \omega \big(|x - x_0| + |u - u_0| \big) \, dx \right)^{\frac{1}{2}} \left(\fint_{B_\varrho(x_0)} |Du|^2 \, dx \right)^{\frac{1}{2}}$$

$$\leq c(L) \omega^{1/2} \left(\varrho + \left(\fint_{B_\varrho(x_0)} |u - u_0|^2 \, dx \right)^{\frac{1}{2}} \right) \left(\fint_{B_\varrho(x_0)} |Du|^2 \, dx \right)^{\frac{1}{2}}. \qquad \square$$

Before proceeding to the alternative proof of the excess decay estimate we recall from (4.15) the definition of the excess

$$E(u; x_0, \varrho) := \fint_{B_\varrho(x_0)} |u - (u)_{B_\varrho(x_0)}|^2 \, dx$$

for every ball $B_\varrho(x_0) \subset \Omega$. We now introduce a similar excess quantity (appearing on the right-hand side in the estimate of Lemma 4.28) via

$$\tilde{E}(u; x_0, \varrho) := \varrho^{2-n} \int_{B_\varrho(x_0)} |Du|^2 \, dx \,.$$

From the Poincaré and the Caccioppoli inequalities (Lemma 1.56 and Proposition 4.5, respectively) we observe that both excess quantities are comparable for weak solutions to (4.13) in the limit $\varrho \searrow 0$, since we have

$$E(u; x_0, \varrho) \le c(n, N)\tilde{E}(u; x_0, \varrho) \le c(n, N, L)E(u; x_0, 2\varrho) \,. \tag{4.24}$$

First alternative proof of Lemma 4.21 Let $B_R(x_0) \subset \Omega$. For what follows, we may suppose that $E(u; x_0, R/2) \ne 0$ holds (which in particular implies $\tilde{E}(u; x_0, R/2) \ne 0$ by Poincaré's inequality), since otherwise the excess decay estimate is trivially satisfied. We start by defining a rescaling $w \in W^{1,2}(B_{R/2}(x_0), \mathbb{R}^N)$ of the weak solution u via

$$w(x) := u(x)(\tilde{E}(u; x_0, R/2))^{-1/2} \,.$$

Thus, we obtain

$$\left(\frac{R}{2}\right)^{2-n} \int_{B_{R/2}(x_0)} |Dw|^2 \, dx \le 1 \,,$$

and Lemma 4.28 with the choice $u_0 := (u)_{B_{R/2}(x_0)}$ gives

$$\left| \left(\frac{R}{2}\right)^{1-n} \int_{B_{R/2}(x_0)} a(x_0, (u)_{B_{R/2}(x_0)})Dw \cdot D\varphi \, dx \right|$$
$$\le c(n, L)\omega^{1/2}\big(R + E(u; x_0, R/2)^{\frac{1}{2}}\big) \sup_{B_{R/2}(x_0)} |D\varphi|$$

for all functions $\varphi \in C_0^1(B_{R/2}(x_0), \mathbb{R}^N)$. Now we consider $\varepsilon > 0$ (to be fixed later) and take $\delta = \delta(n, N, L, \varepsilon) > 0$ to be the constant from the $\mathcal{A}$-harmonic approximation Lemma 4.27. Assuming the smallness condition

$$c(n, L)\omega^{1/2}\big(R + E(u; x_0, R/2)^{\frac{1}{2}}\big) \le \delta \,, \tag{4.25}$$

we deduce that all assumptions of Lemma 4.27 are satisfied for the function w on $B_{R/2}(x_0)$, with $\gamma = 1$ and the bilinear form given by $\mathcal{A} := a(x_0, (u)_{B_{R/2}(x_0)})$, which is elliptic and bounded by L due to (4.6) and (4.7). Consequently, we can find an $\mathcal{A}$-harmonic function $h \in W^{1,2}(B_{R/2}(x_0), \mathbb{R}^N)$ such that

$$\left(\frac{R}{2}\right)^{-n} \int_{B_{R/2}(x_0)} |w - h|^2 \, dx \leq \varepsilon \quad \text{and} \quad \left(\frac{R}{2}\right)^{2-n} \int_{B_{R/2}(x_0)} |Dh|^2 \, dx \leq 1$$
(4.26)

hold. Since h satisfies the decay estimates given in Lemma 4.11, we find for every $\tau \in (0, 1/2)$

$$E(u; x_0, \tau R) = \fint_{B_{\tau R}(x_0)} |u - (u)_{B_{\tau R}(x_0)}|^2 \, dx$$

$$\leq \fint_{B_{\tau R}(x_0)} |u - (\tilde{E}(u; x_0, R/2))^{1/2} (h)_{B_{\tau R}(x_0)}|^2 \, dx$$

$$= \tilde{E}(u; x_0, R/2) \fint_{B_{\tau R}(x_0)} |w - (h)_{B_{\tau R}(x_0)}|^2 \, dx$$

$$\leq 2\tilde{E}(u; x_0, R/2) \left[\fint_{B_{\tau R}(x_0)} |w - h|^2 \, dx + \fint_{B_{\tau R}(x_0)} |h - (h)_{B_{\tau R}(x_0)}|^2 \, dx \right]$$

$$\leq 2\tilde{E}(u; x_0, R/2) \left[(2\tau)^{-n} \fint_{B_{R/2}(x_0)} |w - h|^2 \, dx \right.$$

$$\left. + c(n, N, L)\tau^2 \fint_{B_{R/2}(x_0)} |h - (h)_{B_{R/2}(x_0)}|^2 \, dx \right].$$

Then, using Poincaré's inequality and employing the estimates in (4.26), we arrive at

$$E(u; x_0, \tau R)$$

$$\leq c(n, N, L)\tilde{E}(u; x_0, R/2) \left[\tau^{-n}\varepsilon + \tau^2 \left(\frac{R}{2}\right)^{2-n} \int_{B_{R/2}(x_0)} |Dh|^2 \, dx \right]$$

$$\leq c(n, N, L)\tilde{E}(u; x_0, R/2) \left[\tau^{-n}\varepsilon + \tau^2 \right].$$

With the choice $\varepsilon = \tau^{n+2}$ (which in turn determines δ in dependency of n, N, L, and τ), the Caccioppoli inequality from Proposition 4.5 then establishes the desired excess decay estimate

$$E(u; x_0, \tau R) \leq c(n, N, L)\tau^2 \tilde{E}(u; x_0, R/2) \leq c(n, N, L)\tau^2 E(u; x_0, R)$$

for all $\tau \in (0, 1/2)$ (and $\tau \in [1/2, 1)$ is again trivial), provided that the smallness condition (4.25) is satisfied. This is precisely the excess decay estimate (4.17), and thus it only remains to comment on the validity of (4.25). Clearly, we can determine the parameters ε_0 and R_0 depending only on n, N, L, ω, and τ in such a way that $R \leq R_0$ and $E(u; x_0, R) < \varepsilon_0^2$ guarantee (4.25) (and in turn the claim (4.17)). This finishes the proof of the lemma. $\square$

4.3.3 The Direct Approach

The third and last approach, which implements a comparison principle with a solution to the linearized system, is called the direct method. We now explain the underlying ideas of this technique by considering weak solutions $u \in W^{1,2}(\Omega, \mathbb{R}^N)$ to quasilinear systems of the form (4.13), and we give another proof of the excess decay estimate from Lemma 4.21 in this simple model case. Here we follow the proof of Giaquinta and Giusti [31] (who in fact gave the corresponding result for inhomogeneous systems), but we note that the same method was employed independently by Ivert [48]. The excess decay estimate is now obtained by a direct comparison principle, which is quite similar to the perturbation argument in the derivation of the excess decay estimate for linear systems with x-dependent coefficients, see the proof of Lemma 4.16. In that situation, the original solution was compared to the solution of the frozen system in a suitable Dirichlet class (preserving the boundary values of the weak solution).

Before we proceed to the proof of Lemma 4.21, we first establish a local higher integrability result for Du, which is in fact one of the central ingredients of the direct approach.

Proposition 4.29 *Let $u \in W^{1,2}(\Omega, \mathbb{R}^N)$ be a weak solution to the system (4.13) with Carathéodory coefficients $a\colon \Omega \times \mathbb{R}^N \to \mathbb{R}^{Nn \times Nn}$ which satisfy (4.6) and (4.7). Then there exists a number $p > 2$ depending only on n, N, and L such that we have $u \in W_{\mathrm{loc}}^{1,p}(\Omega, \mathbb{R}^N)$, and for every ball $B_R(x_0) \subset \Omega$ there holds*

$$\left(\fint_{B_{R/2}(x_0)} \left(1 + |Du| \right)^p dx \right)^{\frac{2}{p}} \leq c(n, N, L) \fint_{B_R(x_0)} \left(1 + |Du|^2 \right) dx \, .$$

Proof It suffices to verify the prerequisite of Gehring's Theorem 1.22. To this end we combine the Caccioppoli inequality from Proposition 4.5 with the

Sobolev–Poincaré inequality, see Remark 1.57 (iv), and we find

$$\fint_{B_{\varrho/2}(y)} |Du|^2\,dx \le c(L)\varrho^{-2} \fint_{B_{\varrho}(y)} |u - (u)_{B_{\varrho}(y)}|^2\,dx$$

$$\le c(n,N,L)\varrho^{-2-n}\left(\int_{B_{\varrho}(y)} |Du|^{\frac{2n}{n+2}}\,dx\right)^{\frac{n+2}{n}}$$

$$= c(n,N,L)\left(\fint_{B_{\varrho}(y)} |Du|^{\frac{2n}{n+2}}\,dx\right)^{\frac{n+2}{n}}$$

for all balls $B_{\varrho}(y) \subset \Omega$. Thus, the reverse-Hölder type inequality in the assumptions of Theorem 1.22 is satisfied with $f = |Du|^2$, $g = 0$, $\sigma = 1/2$, and $m = n/(n+2) \in (0,1)$. Therefore, the asserted local higher integrability is justified for an exponent $p > 2$ and a constant c, which both depend only on n, N, and L. □

Second alternative proof of Lemma 4.21 We take $B_R(x_0) \subset \Omega$ and start by considering the Dirichlet problem

$$\begin{cases} \operatorname{div}\big(a(x_0,(u)_{B_{R/4}(x_0)})Dv\big) = 0 & \text{in } B_{R/4}(x_0)\,, \\ v = u & \text{on } \partial B_{R/4}(x_0)\,. \end{cases}$$

Using the Lax–Milgram Theorem A.11 as explained in Remark 4.12, we find a unique weak solution v in the Dirichlet class $u + W_0^{1,2}(B_{R/4}(x_0),\mathbb{R}^N)$. Our first goal is to show that v is close to the original solution provided that x_0 is a regular point and that R is sufficiently small. Indeed, we first observe

$$\int_{B_{R/4}(x_0)} a(x_0,(u)_{B_{R/4}(x_0)})(Du - Dv)\cdot D\varphi\,dx$$

$$= \int_{B_{R/4}(x_0)} \big[a(x_0,(u)_{B_{R/4}(x_0)}) - a(x,u)\big]Du\cdot D\varphi\,dx. \quad (4.27)$$

for all $\varphi \in W_0^{1,2}(B_{R/4}(x_0),\mathbb{R}^N)$. Testing the latter identity with $\varphi = u - v$ and keeping in mind the ellipticity condition (4.6), we get

$$\int_{B_{R/4}(x_0)} |Du - Dv|^2\,dx$$

$$\le \int_{B_{R/4}(x_0)} a(x_0,(u)_{B_{R/4}(x_0)})(Du - Dv)\cdot(Du - Dv)\,dx$$

$$= \int_{B_{R/4}(x_0)} [a(x_0, (u)_{B_{R/4}(x_0)}) - a(x, u)] \, Du \cdot (Du - Dv) \, dx$$

$$\leq c(L) \int_{B_{R/4}(x_0)} \omega^\theta \big(|x - x_0| + |u - (u)_{B_{R/4}(x_0)}| \big) |Du| |Du - Dv| \, dx$$

with $\theta \in [0, 1]$ arbitrary, where we have used the boundedness (4.7) and uniform continuity (4.14) of the coefficients. From Young's inequality we hence obtain

$$\int_{B_{R/4}(x_0)} |Du - Dv|^2 \, dx$$

$$\leq c(L) \int_{B_{R/4}(x_0)} \omega^{2\theta} \big(|x - x_0| + |u - (u)_{B_{R/4}(x_0)}| \big) |Du|^2 \, dx . \quad (4.28)$$

This inequality is not yet sufficient in order to find a good decay estimate for Du since the $\omega^{2\theta}$-factor in the integral in (4.28) is not uniformly small (which means that this term cannot be made arbitrarily small, merely by assuming a small initial radius R and a small initial excess $E(u; x_0, R)$ as claimed). At this stage, the previous higher integrability result comes into play, yielding the reverse Hölder inequality

$$\Big(\fint_{B_{R/4}(x_0)} |Du|^p \, dx \Big)^{\frac{1}{p}} \leq c(n, N, L) \Big(\fint_{B_{R/2}(x_0)} |Du|^2 \, dx \Big)^{\frac{1}{2}} .$$

Choosing $2\theta = (1 - 2/p) \in (0, 1)$ and employing Hölder's and Jensen's inequalities, we then infer for the right-hand side of inequality (4.28)

$$\fint_{B_{R/4}(x_0)} \omega^{2\theta} \big(|x - x_0| + |u - (u)_{B_{R/4}(x_0)}| \big) |Du|^2 \, dx$$

$$\leq \Big(\fint_{B_{R/4}(x_0)} \omega \big(|x - x_0| + |u - (u)_{B_{R/4}(x_0)}| \big) \, dx \Big)^{1 - \frac{2}{p}} \Big(\fint_{B_{R/4}(x_0)} |Du|^p \, dx \Big)^{\frac{2}{p}}$$

$$\leq \omega^{1 - 2/p} \Big(R + \Big(\fint_{B_{R/4}(x_0)} |u - (u)_{B_{R/4}(x_0)}|^2 \, dx \Big)^{\frac{1}{2}} \Big) \Big(\fint_{B_{R/4}(x_0)} |Du|^p \, dx \Big)^{\frac{2}{p}} .$$

Plugging this into (4.28) and taking advantage of the higher integrability estimate, we end up with

$$\int_{B_{R/4}(x_0)} |Du - Dv|^2 \, dx$$

$$\leq c(n, N, L) \omega^{1 - 2/p} \big(R + c(n) E(u; x_0, R)^{1/2} \big) \int_{B_{R/2}(x_0)} |Du|^2 \, dx . \quad (4.29)$$

With this estimate for the L^2-distance between Du and Dv at hand, our next objective is to deduce a Morrey-type decay estimate for the gradient Du (which is essentially equivalent to the Campanato-type decay estimates for u derived before, cf. (4.24), but gradient estimates here appear to be the more natural ones). Since v is a solution of an elliptic system with constant coefficients, the decay estimates from Lemma 4.11 imply for every $\tau \in (0, 1/4)$:

$$\int_{B_{\tau R}(x_0)} |Du|^2\, dx \le 2 \int_{B_{\tau R}(x_0)} |Du - Dv|^2\, dx + 2 \int_{B_{\tau R}(x_0)} |Dv|^2\, dx$$

$$\le 2 \int_{B_{R/4}(x_0)} |Du - Dv|^2\, dx + c(n, N, L)(4\tau)^n \int_{B_{R/4}(x_0)} |Dv|^2\, dx$$

$$\le c(n, N, L) \int_{B_{R/4}(x_0)} |Du - Dv|^2\, dx + c(n, N, L)\tau^n \int_{B_{R/4}(x_0)} |Du|^2\, dx$$

$$\le c(n, N, L)\Big(\omega^{1-2/p}\big(R + c(n)E(u; x_0, R)^{1/2}\big) + \tau^n\Big) \int_{B_{R/2}(x_0)} |Du|^2\, dx\,.$$

Therefore, under the smallness assumption

$$\omega^{1-2/p}\big(R + c(n)E(u; x_0, R)^{1/2}\big) \le \tau^n\,, \tag{4.30}$$

we obtain the inequality

$$\int_{B_{\tau R}(x_0)} |Du|^2\, dx \le c(n, N, L)\tau^n \int_{B_{R/2}(x_0)} |Du|^2\, dx\,,$$

which then, via the Poincaré and Caccioppoli inequalities (Lemma 1.56 and Proposition 4.5), yields the decay estimate

$$E(u; x_0, \tau R) = \fint_{B_{\tau R}(x_0)} |u - (u)_{B_{\tau R}(x_0)}|^2\, dx$$

$$\le c(n, N)(\tau R)^2 \fint_{B_{\tau R}(x_0)} |Du|^2\, dx$$

$$\le c(n, N, L)(\tau R)^2 \fint_{B_{R/2}(x_0)} |Du|^2\, dx$$

$$\le c(n, N, L)\tau^2 \fint_{B_R(x_0)} |u - (u)_{B_R(x_0)}|^2\, dx$$

$$= c(n, N, L)\tau^2 E(u; x_0, R)\,.$$

In conclusion, we arrive at the excess decay estimate stated in Lemma 4.21 provided that the smallness assumption (4.30) on the initial radius R and on the excess on the initial ball $B_R(x_0)$ are satisfied (which in turn gives the asserted dependencies for the constants ε_0 and R_0, by noting that the higher integrability exponent p depends only on n, N, and L).   □

For comparison: A partial regularity result for minimizers to quadratic variational integrals To conclude this chapter on the fundamentals on (partial) regularity for elliptic problems in the vectorial case, we sketch a related result for a simple minimization problem, namely of quadratic functionals of the form

$$\mathcal{Q}[w; \Omega] := \int_\Omega a(x, w) Dw \cdot Dw \, dx$$

with symmetric Carathéodory coefficients $a \colon \Omega \times \mathbb{R}^N \to \mathbb{R}^{Nn \times Nn}$, and we now wish to investigate the regularity of minimizers in Dirichlet classes of $W^{1,2}(\Omega, \mathbb{R}^N)$. In a regular setting the associated Euler–Lagrange system is of the type (4.13), cf. Lemma 2.8, but our main interest lies in the case for which only low regularity of the coefficients is supposed. In particular, if the coefficients are not differentiable with respect to the u-variable, then the Euler–Lagrange system does in general not exist, and consequently, partial regularity is not any more a consequence of the previous results on weak solutions to systems of the form (4.13). However, it is still possible to follow a modified version of any of the previous approaches. For illustration, here we implement a version of the direct approach in order to prove the following partial regularity result for minimizers, due to Giaquinta and Giusti [32].

Theorem 4.30 (Giaquinta and Giusti)   *Let $u \in W^{1,2}(\Omega, \mathbb{R}^N)$ be a minimizer of the functional $\mathcal{Q}$ with continuous coefficients $a \colon \Omega \times \mathbb{R}^N \to \mathbb{R}^{Nn \times Nn}$ satisfying (4.6), (4.7) and (4.14). Then we have the characterization of the singular set via*

$$\mathrm{Sing}_0(u) = \left\{ x_0 \in \Omega \colon \liminf_{\varrho \searrow 0} \fint_{\Omega(x_0, \varrho)} |u - (u)_{\Omega(x_0, \varrho)}|^2 \, dx > 0 \right\}$$

and $\dim_{\mathcal{H}}(\mathrm{Sing}_0(u)) \le n - 2$. Moreover, for every $\alpha \in (0, 1)$ there holds $\mathrm{Reg}_0(u) = \mathrm{Reg}_\alpha(u)$, i.e. $u \in C^{0,\alpha}(\mathrm{Reg}_0(u), \mathbb{R}^N)$.

The starting point is again a suitable version of a Caccioppoli inequality.

Proposition 4.31 (Caccioppoli inequality)   *Let $u \in W^{1,2}(\Omega, \mathbb{R}^N)$ be a minimizer of the functional $\mathcal{Q}$ with Carathéodory coefficients $a \colon \Omega \times \mathbb{R}^N \to \mathbb{R}^{Nn \times Nn}$ satisfying (4.6) and (4.7). Then we have for all $\zeta \in \mathbb{R}^N$ and for all $B_r(x_0) \Subset B_R(x_0) \subset \Omega$*

$$\int_{B_r(x_0)} |Du|^2 \, dx \le c(L)(R - r)^{-2} \int_{B_R(x_0)} |u - \zeta|^2 \, dx \,.$$

Proof We consider $r \leq \varrho < \sigma \leq R$ and take a cut-off function $\eta \in C_0^\infty(B_\sigma(x_0), [0,1])$ satisfying $\eta \equiv 1$ in $B_\varrho(x_0)$ and $|D\eta| \leq 2(\sigma - \varrho)^{-1}$. We now use as a competitor the function

$$u + \varphi := u - \eta(u - \zeta) \quad \in u + W^{1,2}(\Omega, \mathbb{R}^N),$$

which essentially means that we compare the minimizer u to a modification of u obtained by replacing u with the constant ζ in the smaller ball $B_\varrho(x_0)$ (and interpolating u and ζ in the annulus $B_\sigma(x_0) \setminus B_\varrho(x_0)$). Since u coincides with $u + \varphi$ outside of $B_\sigma(x_0)$, we infer from the minimality of u the inequality $\mathcal{Q}[u; B_\sigma(x_0)] \leq \mathcal{Q}[u + \varphi; B_\sigma(x_0)]$, which, via (4.6), (4.7) and Young's inequality, in turn implies

$$\int_{B_\sigma(x_0)} |Du|^2 \, dx \leq \int_{B_\sigma(x_0)} a(x,u) Du \cdot Du \, dx$$

$$\leq \int_{B_\sigma(x_0)} a(x, u + \varphi)(Du + D\varphi) \cdot (Du + D\varphi) \, dx$$

$$\leq L \int_{B_\sigma(x_0)} |Du + D\varphi|^2 \, dx$$

$$\leq c(L)(\sigma - \varrho)^{-2} \int_{B_\sigma(x_0)} |u - \zeta|^2 \, dx + c^*(L) \int_{B_\sigma(x_0) \setminus B_\varrho(x_0)} |Du|^2 \, dx.$$

Adding $c^*(L) \int_{B_\varrho(x_0)} |Du|^2 \, dx$ to both sides (in order to apply the hole-filling technique), we then find

$$\int_{B_\varrho(x_0)} |Du|^2 \, dx \leq c(L)(\sigma - \varrho)^{-2} \int_{B_R(x_0)} |u - \zeta|^2 \, dx$$

$$+ \frac{c^*(L)}{c^*(L) + 1} \int_{B_\sigma(x_0)} |Du|^2 \, dx,$$

and the iteration Lemma B.1 finally yields the assertion. $\qquad\square$

With the help of the Sobolev–Poincaré inequality and Gehring's lemma, we then obtain, analogously to the proof of Proposition 4.29, a (local) higher integrability result.

Proposition 4.32 *Let $u \in W^{1,2}(\Omega, \mathbb{R}^N)$ be a minimizer of the functional $\mathcal{Q}$ with Carathéodory coefficients $a: \Omega \times \mathbb{R}^N \to \mathbb{R}^{Nn \times Nn}$ satisfying (4.6) and (4.7). Then there exists a number $p > 2$ depending only on n, N, and L such that $u \in W^{1,p}_{\mathrm{loc}}(\Omega, \mathbb{R}^N)$, and for every ball $B_R(x_0) \subset \Omega$ there holds*

$$\left(\fint_{B_{R/2}(x_0)} |Du|^p \, dx \right)^{\frac{2}{p}} \leq c(n, N, L) \fint_{B_R(x_0)} |Du|^2 \, dx.$$

With these preliminary estimates at hand, we may now proceed to the proof of the partial regularity result.

Proof of Theorem 4.30 We may follow the line of argument of the proof of the related Theorem 4.23 and Corollary 4.25 for elliptic systems, and hence, it only remains to establish an excess decay estimate as in Lemma 4.21. For this purpose, we follow the strategy of the second alternative proof of Lemma 4.21, that is, we use the direct approach, up to some modifications which are necessary since we now work with variational integrals and not any more with elliptic systems. We start by considering, for $B_R(x_0) \subset \Omega$, the unique function $v \in u + W_0^{1,2}(\Omega, \mathbb{R}^N)$ which is a weak solution to the system

$$\operatorname{div}\big(a(x_0, (u)_{B_{R/4}(x_0)})Dv\big) = 0 \qquad \text{in } B_{R/4}(x_0) \qquad (4.31)$$

(note that existence is guaranteed by Lax–Milgram Theorem A.11) or equivalently, which minimizes the functional

$$\int_{B_{R/4}(x_0)} a(x_0, (u)_{B_{R/4}(x_0)})Dw \cdot Dw \, dx$$

among all functions $w \in u + W_0^{1,2}(\Omega, \mathbb{R}^N)$ (the equivalence of these two formulations is a consequence of the convexity of the integrand, which implies that every weak solution of the Euler–Lagrange system is in fact a minimizer). Concerning the function v we have on the one hand the excess decay estimates from Lemma 4.11 at our disposal, but on the other hand Dv and Du are also comparable in the $L^2(B_{R/4}(x_0), \mathbb{R}^{Nn})$-sense. In fact, via (4.6) and (4.7) we find

$$\int_{B_{R/4}(x_0)} |Dv|^2 \, dx \le \int_{B_{R/4}(x_0)} a(x_0, (u)_{B_{R/4}(x_0)})Dv \cdot Dv \, dx$$

$$\le \int_{B_{R/4}(x_0)} a(x_0, (u)_{B_{R/4}(x_0)})Du \cdot Du \, dx$$

$$\le L \int_{B_{R/4}(x_0)} |Du|^2 \, dx \qquad (4.32)$$

(and the same is true with interchanged roles of u and v if we consider the original minimization problem). Moreover, in order to estimates the L^2-distance between Du and Dv, we first observe from (4.6) and the weak formulation of the system (4.31)

$$\int_{B_{R/4}(x_0)} |Du - Dv|^2 \, dx$$

$$\le \int_{B_{R/4}(x_0)} a(x_0, (u)_{B_{R/4}(x_0)})(Du - Dv) \cdot (Du - Dv) \, dx$$

$$= \int_{B_{R/4}(x_0)} a(x_0,(u)_{B_{R/4}(x_0)})Du \cdot (Du - Dv)\,dx$$

$$= \int_{B_{R/4}(x_0)} \left[a(x_0,(u)_{B_{R/4}(x_0)}) - a(x,u)\right](Du + Dv) \cdot (Du - Dv)\,dx$$

$$+ \int_{B_{R/4}(x_0)} a(x,u)Du \cdot Du\,dx - \int_{B_{R/4}(x_0)} a(x,v)Dv \cdot Dv\,dx$$

$$+ \int_{B_{R/4}(x_0)} \left[a(x,v) - a(x,u)\right]Dv \cdot Dv\,dx\,.$$

In view of the minimality of u the second-last line is non-positive, and therefore, (4.14) and Young's inequality allow us to conclude that

$$\int_{B_{R/4}(x_0)} |Du - Dv|^2\,dx$$

$$\leq \int_{B_{R/4}(x_0)} \left[|Du|^2 + |Dv|^2\right]\omega\left(|x - x_0| + |u - (u)_{B_{R/4}(x_0)}| + |u - v|\right)dx\,.$$

Next, we observe that also Dv has higher integrability, without loss of generality with the same exponent $p > 2$ (depending only on n, N, and L) as Du, and it satisfies the estimate

$$\left(\fint_{B_{R/4}(x_0)} |Dv|^p\,dx\right)^{\frac{2}{p}} \leq c(n,N,L) \fint_{B_{R/2}(x_0)} |Du|^2\,dx\,.$$

This follows as a consequence of the so-called L^p-theory (combined with the higher integrability estimate for Du), but alternatively we may interpret $v - u \in W_0^{1,2}(B_{R/4}(x_0),\mathbb{R}^N)$ as a weak solution of a system of the form (4.5) (with inhomogeneities given by $f = -a(x_0,(u)_{B_{R/4}(x_0)})Du$ and $g \equiv 0$). Therefore, $v - u$ satisfies the Caccioppoli inequality in Proposition 4.6 and consequently it has also higher integrability by the direct application of Gehring's Theorem 1.22. We next observe

$$\fint_{B_{R/4}(x_0)} |u - v|^2\,dx \leq c(L) \fint_{B_{R/2}(x_0)} |u - (u)_{B_{R/2}(x_0)}|^2\,dx\,,$$

by applying first Poincaré's inequality, then employing (4.32) and finally the Caccioppoli inequality with $\zeta := (u)_{B_{R/2}(x_0)}$. At this stage we may proceed, exactly as for the derivation of (4.29) for elliptic systems, via Hölder's and

Jensen's inequalities, and we find the desired estimate for the L^2-difference of Du and Dv:

$$\int_{B_{R/4}(x_0)} |Du - Dv|^2 \, dx$$

$$\leq c(n, N, L) \omega^{1-2/p} \left(R + cE(u; x_0, R)^{\frac{1}{2}} \right) \int_{B_{R/2}(x_0)} |Du|^2 \, dx \, . \quad (4.33)$$

Now we take $\tau \in (0, 1/4)$ and deduce from the Morrey-type excess decay estimate for Dv in Lemma 4.11, combined with (4.33), the estimate

$$\int_{B_{\tau R}(x_0)} |Du|^2 \, dx \leq 2 \int_{B_{\tau R}(x_0)} |Du - Dv|^2 \, dx + 2 \int_{B_{\tau R}(x_0)} |Dv|^2 \, dx$$

$$\leq 2 \int_{B_{R/4}(x_0)} |Du - Dv|^2 \, dx + c(n, N, L)(4\tau)^n \int_{B_{R/4}(x_0)} |Dv|^2 \, dx$$

$$\leq c(n, N, L) \left(\omega^{1-2/p} \left(R + cE(u; x_0, R)^{\frac{1}{2}} \right) + \tau^n \right) \int_{B_{R/4}(x_0)} |Du|^2 \, dx \, .$$

Under the same smallness assumption $\omega^{1-2/p}(R + cE(u; x_0, R)^{\frac{1}{2}}) \leq \tau^n$ as in (4.30) in the setting of elliptic systems (which requires smallness of the initial radius R and of the initial excess $E(u; x_0, R)$ in terms of n, N, L, ω, and τ), we then obtain

$$\int_{B_{\tau R}(x_0)} |Du|^2 \, dx \leq c(n, N, L) \tau^n \int_{B_{R/2}(x_0)} |Du|^2 \, dx \, ,$$

which, via Poincaré's and of Caccioppoli's inequalities, yields the desired excess decay estimate

$$E(u; x_0, \tau R) \leq c(n, N, L) \tau^2 E(u; x_0, R) \, .$$

This finishes the proof of the corresponding version of Lemma 4.21 for minimizers, and thus the proof of the partial regularity result is complete. $\quad \square$

Chapter 5
Partial Regularity Results for Quasilinear Systems

We next study more general quasilinear systems in divergence form

$$\operatorname{div} a(x, u, Du) = 0 \quad \text{in } \Omega \tag{5.1}$$

and deal with partial $C^{1,\alpha}$-regularity results for their weak solutions. More precisely, we first give a basic higher integrability statement. Then we employ the method of $\mathcal{A}$-harmonic approximation, which was introduced in the previous chapter, in order to prove in the first place the partial C^1-regularity of weak solutions outside of a singular set which is of $\mathcal{L}^n$-measure zero and in the second place the optimal regularity improvement from C^1 to $C^{1,\alpha}$ for some $\alpha > 0$ (determined by the regularity of the vector field a). These results come along with a characterization of the exceptional set on which singularities of a weak solution may arise. However, it does not directly allow for a non-trivial bound on its Hausdorff dimensions, but this requires further work. In different settings, from simple to quite general ones, we explain (fractional) higher differentiability estimates for the gradient of weak solutions. These provide, in turn, the desired bounds for the Hausdorff dimension of the singular set.

5.1 Initial Observations and Higher Integrability

Throughout this chapter, we impose on the Carathéodory vector field $a \colon \Omega \times \mathbb{R}^N \times \mathbb{R}^{Nn} \to \mathbb{R}^{Nn}$ the following assumptions concerning growth, differentiability, and ellipticity:

(H1) a is differentiable with respect to z with

$$|a(x, u, z)| + |D_z a(x, u, z)|(1 + |z|) \leq L(1 + |z|),$$

© Springer International Publishing Switzerland 2016
L. Beck, *Elliptic Regularity Theory*, Lecture Notes of the Unione
Matematica Italiana 19, DOI 10.1007/978-3-319-27485-0_5

(H2) a is uniformly elliptic in the sense of

$$D_z a(x, u, z)\xi \cdot \xi \geq |\xi|^2 \qquad \text{for all } \xi \in \mathbb{R}^{Nn},$$

for almost every $x \in \Omega$, all $(u, z) \in \mathbb{R}^N \times \mathbb{R}^{Nn}$, and with a constant $L \geq 1$. We note that these assumptions essentially guarantee that the vector field a is possibly nonlinear in the gradient variable, but still of linear growth (see also the Remarks 5.1 below), and the particular quasilinear system considered in the previous chapter satisfies in particular these assumptions. Later on, we will assume further continuity assumptions on the vector field a, but we here want to comment on some general facts, which rely only on the assumptions (H1) and (H2).

Remarks 5.1

(i) The assumption (H1) implies $a(\cdot, u, Du) \in L^2(\Omega, \mathbb{R}^{Nn})$ for every $u \in W^{1,2}(\Omega, \mathbb{R}^N)$. Consequently, via an approximation argument (see also Remark 2.5), a function $u \in W^{1,2}(\Omega, \mathbb{R}^N)$ is a weak solution to the system (5.1) if and only if we have

$$\int_\Omega a(x, u, Du) \cdot D\varphi\, dx = 0 \qquad \text{for all } \varphi \in W_0^{1,2}(\Omega, \mathbb{R}^N). \qquad (5.2)$$

(ii) For almost every $x \in \Omega$, all $u \in \mathbb{R}^N$, and all $z, \tilde{z} \in \mathbb{R}^{Nn}$ we observe from the bound on $D_z a$ in (H1) that a is Lipschitz continuous with respect to the z-variable, since we have

$$|a(x, u, z) - a(x, u, \tilde{z})| = \left| \int_0^1 \frac{d}{dt} a(x, u, tz + (1-t)\tilde{z})\, dt \right|$$

$$= \left| \int_0^1 D_z a(x, u, tz + (1-t)\tilde{z})\, dt\, (z - \tilde{z}) \right|$$

$$\leq L|z - \tilde{z}|.$$

(iii) For almost every $x \in \Omega$, all $u \in \mathbb{R}^N$, and all $z, \tilde{z} \in \mathbb{R}^{Nn}$ the ellipticity condition (H2) gives

$$\big(a(x, u, z) - a(x, u, \tilde{z})\big) \cdot (z - \tilde{z})$$

$$= \int_0^1 D_z a(x, u, tz + (1-t)\tilde{z})\, dt\, (z - \tilde{z}) \cdot (z - \tilde{z}) \geq |z - \tilde{z}|^2$$

(for this reason a is called a monotone operator with respect to the gradient variable).

We next state a very simple version of a Caccioppoli inequality, which is a straightforward extension of Proposition 4.5 from the linear to the quasilinear case.

Proposition 5.2 (Basic Caccioppoli inequality) *Let $u \in W^{1,2}(\Omega, \mathbb{R}^N)$ be a weak solution to the system (5.1) with a Carathéodory vector field $a \colon \Omega \times \mathbb{R}^N \times \mathbb{R}^{Nn} \to \mathbb{R}^{Nn}$ satisfying (H1)–(H2). Then, for all $\zeta \in \mathbb{R}^N$ and all balls $B_r(x_0) \subset \Omega$, we have*

$$\int_{B_{r/2}(x_0)} |Du|^2 \, dx \le c(L) \int_{B_r(x_0)} \left(1 + r^{-2}|u - \zeta|^2\right) dx.$$

Proof We take a cut-off function $\eta \in C_0^\infty(B_r(x_0), [0,1])$ which satisfies $\eta \equiv 1$ in $B_{r/2}(x_0)$ and $|D\eta| \le 4r^{-1}$. We now test the weak formulation (5.2) with the function $\varphi := \eta^2(u - \zeta) \in W_0^{1,2}(\Omega, \mathbb{R}^N)$. In view of the ellipticity estimate in Remark 5.1 (iii) and with (H1), we then find

$$\int_{B_r(x_0)} |Du|^2 \eta^2 \, dx \le \int_{B_r(x_0)} \left(a(x, u, Du) - a(x, u, 0)\right) \cdot Du\eta^2 \, dx$$

$$= -\int_{B_r(x_0)} \left(2a(x, u, Du) \cdot (u - \zeta) \otimes D\eta + a(x, u, 0) \cdot Du\eta\right)\eta \, dx$$

$$\le L \int_{B_r(x_0)} \left(2(1 + |Du|)|u - \zeta||D\eta| + |Du|\right)\eta \, dx,$$

and the assertion now follows from Young's inequality and the choice of η. $\square$

With the previous proposition at hand, a higher integrability of Du, which is analogous to the result of Proposition 4.29 for the linear model systems, follows immediately:

Lemma 5.3 *Let $u \in W^{1,2}(\Omega, \mathbb{R}^N)$ be a weak solution to the system (5.1) with a Carathéodory vector field $a \colon \Omega \times \mathbb{R}^N \times \mathbb{R}^{Nn} \to \mathbb{R}^{Nn}$ satisfying (H1)–(H2). Then there exists a number $p > 2$ depending only on n, N, and L such that $u \in W^{1,p}_{\mathrm{loc}}(\Omega, \mathbb{R}^N)$, and for every ball $B_{2R}(x_0) \subset \Omega$ there holds*

$$\left(\fint_{B_R(x_0)} \left(1 + |Du|\right)^p dx\right)^{\frac{1}{p}} \le c(n, N, L)\left(\fint_{B_{2R}(x_0)} \left(1 + |Du|^2\right) dx\right)^{\frac{1}{2}}. \quad (5.3)$$

Proof With the Sobolev–Poincaré inequality, see Remark 1.57 (iv), we first deduce from Proposition 5.2 (on balls $B_\varrho(y)$ and with the choice $\zeta := (u)_{B_\varrho(y)}$) the reverse Hölder inequalities

$$\fint_{B_{\varrho/2}(y)} \left(1 + |Du|^2\right) dx \le c(n, N, L)\left(\fint_{B_\varrho(y)} \left(1 + |Du|^{\frac{2n}{n+2}}\right) dx\right)^{\frac{n+2}{n}}$$

for all interior balls $B_\varrho(y) \subset \Omega$. The claim now follows from Gehring's Theorem 1.22. □

5.2 Partial $C^{1,\alpha}$-Regularity via the Method of $\mathcal{A}$-Harmonic Approximation

In what follows, we adjust the method of $\mathcal{A}$-harmonic approximation, which was detailed in Sect. 4.3.2, in order to investigate systems of the form (5.1), with the aim to establish a partial C^0-regularity result for the *gradient* of weak solutions (and to prove in addition the optimal regularity improvement to Hölder continuity). For this purpose, we assume for the vector field $a\colon \Omega \times \mathbb{R}^N \times \mathbb{R}^{Nn} \to \mathbb{R}^{Nn}$ in addition to (H1) and (H2) also the following continuity assumptions:

(H3) a is Hölder continuous with respect to (x, u) with

$$|a(x, u, z) - a(\tilde{x}, \tilde{u}, z)| \le 2L\omega_\alpha\big(|x - \tilde{x}| + |u - \tilde{u}|\big)\big(1 + |z|\big),$$

(H4) $D_z a$ is continuous with

$$|D_z a(x, u, z) - D_z a(x, u, \tilde{z})| \le 2L\tilde{\omega}\big(|z - \tilde{z}|\big),$$

for all $x, \tilde{x} \in \Omega$, $u, \tilde{u} \in \mathbb{R}^N$ and $z, \tilde{z} \in \mathbb{R}^{Nn}$. Here, $\omega_\alpha, \tilde{\omega}\colon \mathbb{R}_0^+ \to \mathbb{R}_0^+$ are two moduli of continuity, bounded by 1 from above, monotonically non-decreasing and concave. Furthermore, we suppose $\omega_\alpha(t) \le \min\{1, t^\alpha\}$ for some $\alpha \in (0, 1)$ and $\lim_{t \searrow 0} \tilde{\omega}(t) = \tilde{\omega}(0) = 0$. We note that these assumptions are still satisfied by the particular quasilinear system considered in the previous chapter.

For the proof of the partial $C^{1,\alpha}$-regularity for weak solutions to the system (5.1), we essentially follow the exposition in the paper [21] of Duzaar and Grotowski. The starting point for the investigation of such regularity properties is again a Caccioppoli-type inequality, which now, in contrast to the basic Caccioppoli inequality derived in Proposition 5.2, concerns *affine perturbations* of the weak solution.

Proposition 5.4 (Caccioppoli inequality) *Let* $u \in W^{1,2}(\Omega, \mathbb{R}^N)$ *be a weak solution to the system (5.1) with a vector field* $a\colon \Omega \times \mathbb{R}^N \times \mathbb{R}^{Nn} \to \mathbb{R}^{Nn}$ *satisfying (H1), (H2), and (H3). Then, for all* $\zeta \in \mathbb{R}^N$, $z_0 \in \mathbb{R}^{Nn}$ *and* $B_r(x_0) \subset \Omega$ *with* $r \le 1$, *we have*

$$\int_{B_{r/2}(x_0)} |Du - z_0|^2 \, dx \le c(L) r^{-2} \int_{B_r(x_0)} |u - \zeta - z_0(x - x_0)|^2 \, dx$$

$$+ c(L)(1 + |z_0|)^{\frac{2}{1-\alpha}} r^{n+2\alpha}.$$

Proof We take a standard cut-off function $\eta \in C_0^\infty(B_r(x_0),[0,1])$ which satisfies $\eta \equiv 1$ in $B_{r/2}(x_0)$ and $|D\eta| \leq 4r^{-1}$. We now test the weak formulation of the system (5.1), see (5.2), with the function $\varphi := \eta^2(u - \zeta - z_0(x - x_0)) \in W_0^{1,2}(\Omega, \mathbb{R}^N)$. Thus, we get

$$\int_{B_r(x_0)} a(x,u,Du) \cdot (Du - z_0)\eta^2 \, dx$$

$$= -2 \int_{B_r(x_0)} a(x,u,Du) \cdot (u - \zeta - z_0(x - x_0)) \otimes D\eta\eta \, dx \, .$$

Moreover, we observe

$$-\int_{B_r(x_0)} a(x,u,z_0) \cdot (Du - z_0)\eta^2 \, dx$$

$$= 2 \int_{B_r(x_0)} a(x,u,z_0) \cdot (u - \zeta - z_0(x - x_0)) \otimes D\eta\eta \, dx$$

$$- \int_{B_r(x_0)} a(x,u,z_0) \cdot D\varphi \, dx \, ,$$

and, since $a(x_0,\zeta,z_0)$ is constant, we also have

$$0 = \int_{B_r(x_0)} a(x_0,\zeta,z_0) \cdot D\varphi \, dx \, .$$

We now want to employ the ellipticity (or monotonicity) condition for $Du - z_0$ from Remark 5.1 (iii). For this purpose, we add the three previous identities and find

$$\int_{B_r(x_0)} |Du - z_0|^2\eta^2 \, dx$$

$$\leq \int_{B_r(x_0)} \big(a(x,u,Du) - a(x,u,z_0)\big) \cdot (Du - z_0)\eta^2 \, dx$$

$$= -2 \int_{B_r(x_0)} \big(a(x,u,Du) - a(x,u,z_0)\big) \cdot (u - \zeta - z_0(x - x_0)) \otimes D\eta\eta \, dx$$

$$- \int_{B_r(x_0)} \big(a(x,u,z_0) - a(x,\zeta + z_0(x - x_0),z_0)\big) \cdot D\varphi \, dx$$

$$- \int_{B_r(x_0)} \big(a(x,\zeta + z_0(x - x_0),z_0) - a(x_0,\zeta,z_0)\big) \cdot D\varphi \, dx$$

$$=: I + II + III \tag{5.4}$$

with the obvious abbreviations. The first term I is estimated via Remark 5.1 (ii) and Young's inequality by

$$|I| \le 2L \int_{B_r(x_0)} |Du - z_0||u - \zeta - z_0(x - x_0)||D\eta|\eta \, dx$$

$$\le \frac{1}{4} \int_{B_r(x_0)} |Du - z_0|^2 \eta^2 \, dx + c(L) r^{-2} \int_{B_r(x_0)} |u - \zeta - z_0(x - x_0)|^2 \, dx \, .$$

For the second term II we take advantage of the continuity assumption (H3) (note that ω_α is a Hölder modulus with $\omega_\alpha \le \min\{1, t^\alpha\}$) and apply repeatedly Young's inequality. In this way we find

$$|II| \le 2L(1 + |z_0|) \int_{B_r(x_0)} \omega_\alpha(|u - \zeta - z_0(x - x_0)|)|D\varphi| \, dx$$

$$\le 2L(1 + |z_0|) \int_{B_r(x_0)} |u - \zeta - z_0(x - x_0)|^\alpha$$

$$\times \left(|Du - z_0|\eta^2 + 2|u - \zeta - z_0(x - x_0)||D\eta|\eta \right) dx$$

$$\le \frac{1}{4} \int_{B_r(x_0)} |Du - z_0|^2 \eta^2 \, dx + r^{-2} \int_{B_r(x_0)} |u - \zeta - z_0(x - x_0)|^2 \, dx$$

$$+ c(L)(1 + |z_0|)^2 \int_{B_r(x_0)} |u - \zeta - z_0(x - x_0)|^{2\alpha} r^{-2\alpha} r^{2\alpha} \, dx$$

$$\le \frac{1}{4} \int_{B_r(x_0)} |Du - z_0|^2 \eta^2 \, dx + r^{-2} \int_{B_r(x_0)} |u - \zeta - z_0(x - x_0)|^2 \, dx$$

$$+ c(L)(1 + |z_0|)^{\frac{2}{1-\alpha}} r^{n + \frac{2\alpha}{1-\alpha}} \, .$$

Similarly, we obtain for the last term III

$$|III| \le 2L(1 + |z_0|) \int_{B_r(x_0)} \omega_\alpha((1 + |z_0|)r)|D\varphi| \, dx$$

$$\le \frac{1}{4} \int_{B_r(x_0)} |Du - z_0|^2 \eta^2 \, dx + r^{-2} \int_{B_r(x_0)} |u - \zeta - z_0(x - x_0)|^2 \, dx$$

$$+ c(L)(1 + |z_0|)^{2(1+\alpha)} r^{n+2\alpha} \, .$$

Combining the previous three estimates with (5.4), we thus infer

$$\int_{B_r(x_0)} |Du - z_0|^2 \eta^2 \, dx$$

$$\leq \frac{3}{4} \int_{B_r(x_0)} |Du - z_0|^2 \eta^2 \, dx + c(L)r^{-2} \int_{B_r(x_0)} |u - \zeta - z_0(x - x_0)|^2 \, dx$$

$$+ c(L)\left((1 + |z_0|)^{\frac{2}{1-\alpha}} r^{n+\frac{2\alpha}{1-\alpha}} + (1 + |z_0|)^{2(1+\alpha)} r^{n+2\alpha}\right)$$

$$\leq \frac{3}{4} \int_{B_r(x_0)} |Du - z_0|^2 \eta^2 \, dx + c(L)r^{-2} \int_{B_r(x_0)} |u - \zeta - z_0(x - x_0)|^2 \, dx$$

$$+ c(L)(1 + |z_0|)^{\frac{2}{1-\alpha}} r^{n+2\alpha}$$

(here we have used the assumption $r \leq 1$). The assertion now follows from the choice of cut-off function η, which satisfies $\eta = 1$ in $B_{r/2}(x_0)$.   □

The second step in the partial regularity proof via the approach of $\mathcal{A}$-harmonic approximation is to determine the setting in which the $\mathcal{A}$-harmonic approximation lemma shall be applied. We now show that, for a suitable linearization of the system (which then defines the bilinear form $\mathcal{A}$) and under some appropriate smallness assumption, a rescaling of the weak solution is approximately $\mathcal{A}$-harmonic.

Lemma 5.5 (Approximate $\mathcal{A}$-harmonicity II) *Let $u \in W^{1,2}(\Omega, \mathbb{R}^N)$ be a weak solution to the system (5.1) with a vector field $a \colon \Omega \times \mathbb{R}^N \times \mathbb{R}^{Nn} \to \mathbb{R}^{Nn}$ satisfying (H1), (H2), (H3), and (H4). Then, for all $z_0 \in \mathbb{R}^{Nn}$ and every $B_\varrho(x_0) \subset \Omega$ with $\varrho \leq 1$, we have*

$$\left| \varrho^{-n} \int_{B_\varrho(x_0)} D_z a(x_0, (u)_{B_\varrho(x_0)}, z_0)(Du - z_0) \cdot D\varphi \, dx \right|$$

$$\leq c_{Ap}(n, N, L)\left[\tilde{\omega}^{1/2}\left(\left(\fint_{B_\varrho(x_0)} |Du - z_0|^2 \, dx\right)^{\frac{1}{2}}\right)\left(\fint_{B_\varrho(x_0)} |Du - z_0|^2 \, dx\right)^{\frac{1}{2}}\right.$$

$$\left. + \fint_{B_\varrho(x_0)} |Du - z_0|^2 \, dx + (1 + |z_0|)^{1+\alpha} \varrho^\alpha\right] \sup_{B_\varrho(x_0)} |D\varphi|$$

for all $\varphi \in C_0^1(B_\varrho(x_0), \mathbb{R}^N)$.

Proof We consider an arbitrary function $\varphi \in C_0^1(B_\varrho(x_0), \mathbb{R}^N)$ satisfying $\sup_{B_\varrho(x_0)} |D\varphi| \leq 1$ (the general result then follows after rescaling). Since $a(x_0, (u)_{B_\varrho(x_0)}, z_0)$ is constant, we observe

$$\operatorname{div} a(x_0, (u)_{B_\varrho(x_0)}, z_0) = 0,$$

and therefore, we find

$$\fint_{B_\varrho(x_0)} D_z a(x_0, (u)_{B_\varrho(x_0)}, z_0)(Du - z_0) \cdot D\varphi \, dx$$

$$= \fint_{B_\varrho(x_0)} \int_0^1 \big[D_z a(x_0, (u)_{B_\varrho(x_0)}, z_0)$$

$$- D_z a(x_0, (u)_{B_\varrho(x_0)}, z_0 + t(Du - z_0)) \big] \, dt (Du - z_0) \cdot D\varphi \, dx$$

$$+ \fint_{B_\varrho(x_0)} \big[a(x_0, (u)_{B_\varrho(x_0)}, Du) - a(x_0, (u)_{B_\varrho(x_0)}, z_0) \big] \cdot D\varphi \, dx$$

$$= \fint_{B_\varrho(x_0)} \int_0^1 \big[D_z a(x_0, (u)_{B_\varrho(x_0)}, z_0)$$

$$- D_z a(x_0, (u)_{B_\varrho(x_0)}, z_0 + t(Du - z_0)) \big] \, dt (Du - z_0) \cdot D\varphi \, dx$$

$$+ \fint_{B_\varrho(x_0)} \big[a(x_0, (u)_{B_\varrho(x_0)}, Du) - a(x, u, Du) \big] \cdot D\varphi \, dx .$$

With the continuity assumptions (H3) on $D_z a$ with respect to the gradient variable z and (H4) on a with respect to the first and second argument (x, u) – combined with the boundedness of $\tilde{\omega}$ from above by 1 –, the previous identity yields the estimate

$$\left| \fint_{B_\varrho(x_0)} D_z a(x_0, u_0, z_0)(Du - z_0) \cdot D\varphi \, dx \right| \leq I + II$$

with the abbreviations

$$I := 2L \fint_{B_\varrho(x_0)} \tilde{\omega}^{1/2} (|Du - z_0|) |Du - z_0| \, dx ,$$

$$II := 2L \fint_{B_\varrho(x_0)} \omega_\alpha \big(|x - x_0| + |u - (u)_{B_\varrho(x_0)}| \big) \big(1 + |Du| \big) \, dx .$$

From Hölder's and Jensen's inequality (which is applicable since $\tilde{\omega}$ is concave), we find

$$I \leq 2L \Big(\fint_{B_\varrho(x_0)} \tilde{\omega}(|Du - z_0|) \, dx \Big)^{\frac{1}{2}} \Big(\fint_{B_\varrho(x_0)} |Du - z_0|^2 \, dx \Big)^{\frac{1}{2}}$$

$$\leq 2L \tilde{\omega}^{1/2} \Big(\big(\fint_{B_\varrho(x_0)} |Du - z_0|^2 \, dx \big)^{\frac{1}{2}} \Big) \Big(\fint_{B_\varrho(x_0)} |Du - z_0|^2 \, dx \Big)^{\frac{1}{2}} .$$

For the second term II we use the particular form of ω_α and employ repeatedly Young's inequality. This gives

$$II \le 2L \fint_{B_\varrho(x_0)} \left[\varrho^\alpha + |u - (u)_{B_\varrho(x_0)}|^\alpha\right] (1 + |Du|) \, dx$$

$$\le 2L \fint_{B_\varrho(x_0)} \left[(1 + |z_0|^\alpha)\varrho^\alpha + |u - (u)_{B_\varrho(x_0)} - z_0(x - x_0)|^\alpha\right]$$

$$\times \left(1 + |z_0| + |Du - z_0|\right) dx$$

$$\le cL(1 + |z_0|)^{1+\alpha}\varrho^\alpha + cL(1 + |z_0|)^{2\alpha}\varrho^{2\alpha} + cL \fint_{B_\varrho(x_0)} |Du - z_0|^2 \, dx$$

$$+ L(1 + |z_0|) \fint_{B_\varrho(x_0)} |u - (u)_{B_\varrho(x_0)} - z_0(x - x_0)|^\alpha \, dx$$

$$+ cL \fint_{B_\varrho(x_0)} |u - (u)_{B_\varrho(x_0)} - z_0(x - x_0)|^{2\alpha} \, dx$$

$$\le cL(1 + |z_0|)^{1+\alpha}\varrho^\alpha + cL(1 + |z_0|)^{2\alpha}\varrho^{2\alpha} + cL \fint_{B_\varrho(x_0)} |Du - z_0|^2 \, dx$$

$$+ cL\varrho^{-2} \fint_{B_\varrho(x_0)} |u - (u)_{B_\varrho(x_0)} - z_0(x - x_0)|^2 \, dx$$

$$+ cL(1 + |z_0|)^{\frac{2}{2-\alpha}} \varrho^{\frac{2\alpha}{2-\alpha}} + cL\varrho^{\frac{2\alpha}{1-\alpha}} .$$

With $\varrho \le 1$ (which allows to estimate all terms involving only powers of ϱ and of $1 + |z_0|$ by only one term) and Poincaré's inequality (note that the function $u - (u)_{B_\varrho(x_0)} - z_0(x - x_0)$ has vanishing mean value on $B_\varrho(x_0)$) we then obtain

$$II \le c(n, N)L \fint_{B_\varrho(x_0)} |Du - z_0|^2 \, dx + cL(1 + |z_0|)^{1+\alpha}\varrho^\alpha .$$

Inserting the estimates for I and II above, we finally arrive at the claim of the lemma. $\qquad\square$

With the Caccioppoli inequality and the approximate $\mathcal{A}$-harmonicity lemma at hand, we can now proceed to a *preliminary excess decay estimate*.

Lemma 5.6 *Let $u \in W^{1,2}(\Omega, \mathbb{R}^N)$ be a weak solution to the system (5.1) with a vector field $a: \Omega \times \mathbb{R}^N \times \mathbb{R}^{Nn} \to \mathbb{R}^{Nn}$ satisfying* (H1), (H2), (H3), *and* (H4). *Then there exist two parameters $\tau = \tau(n, N, L, \alpha) \in (0,1)$ and $\tilde{\varepsilon}_0 = \tilde{\varepsilon}_0(n, N, L, \alpha, \tilde{\omega}) \in (0,1)$ such that*

$$E(Du; x_0, \tau R) \le \tau^{1+\alpha} E(Du; x_0, R) + c_{Dec}(n, N, L, \alpha, M_0)(\tau R)^{2\alpha} \qquad (5.5)$$

holds for every ball $B_R(x_0) \subset \Omega$ with radius $R \leq 1$, provided that we have

$$E(Du; x_0, R) \leq 2\bar{\varepsilon}_0^2 \qquad and \qquad |(Du)_{B_R(x_0)}| \leq 2M_0 \,.$$

Proof For $\varepsilon > 0$ to be determined later we take $\delta = \delta(n, N, L, \varepsilon) > 0$ according to the $\mathcal{A}$-harmonic approximation Lemma 4.27. Then, motivated from the estimate in the previous Lemma 5.5, we define a bilinear form $\mathcal{A} \in \mathbb{R}^{Nn \times Nn}$ and a rescaling of the weak solution u via

$$\mathcal{A} := D_z a(x_0, (u)_{B_R(x_0)}, (Du)_{B_R(x_0)}) \,,$$

$$\gamma := c_{Ap} \big[E(Du; x_0, R) + 4\delta^{-2}(1 + |(Du)_{B_R(x_0)}|)^{2(1+\alpha)} R^{2\alpha} \big]^{\frac{1}{2}}, \qquad (5.6)$$

$$w(x) := \gamma^{-1} \big[u(x) - (Du)_{B_R(x_0)}(x - x_0) \big] \,,$$

where $B_R(x_0) \subset \Omega$ is a ball of radius $R \leq 1$, c_{Ap} is the constant from Lemma 5.5 and the excess $E(Du; x_0, R)$ is defined as in (4.15). Due to the assumptions (H1) and (H2) we observe that $\mathcal{A}$ is bounded and elliptic, and we easily verify (noting that it is not restrictive to assume $c_{Ap}^2 \geq \mathcal{L}^n(B_1)$)

$$R^{-n} \int_{B_R(x_0)} |Dw|^2 \, dx \leq 1 \,.$$

Moreover, Lemma 5.5 (with the choice $z_0 = (Du)_{B_R(x_0)}$) implies that w is approximate $\mathcal{A}$-harmonic with the estimate

$$\left| R^{-n} \int_{B_R(x_0)} \mathcal{A}Dw \cdot D\varphi \, dx \right|$$

$$\leq \frac{\tilde{\omega}^{1/2} \big(E^{1/2}(Du; x_0, R) \big) E^{1/2}(Du; x_0, R) + E(Du; x_0, R)}{E^{1/2}(Du; x_0, R)} \sup_{B_R(x_0)} |D\varphi|$$

$$+ \frac{(1 + |(Du)_{B_R(x_0)}|)^{1+\alpha} R^\alpha}{\big[4\delta^{-2}(1 + |(Du)_{B_R(x_0)}|)^{2(1+\alpha)} R^{2\alpha} \big]^{1/2}} \sup_{B_R(x_0)} |D\varphi|$$

$$\leq \Big[\tilde{\omega}^{1/2} \big(E^{1/2}(Du; x_0, R) \big) + E^{1/2}(Du; x_0, R) + \frac{\delta}{2} \Big] \sup_{B_R(x_0)} |D\varphi| \,.$$

If the smallness assumption

$$\tilde{\omega}^{1/2} \big(E^{1/2}(Du; x_0, R) \big) + E^{1/2}(Du; x_0, R) \leq \frac{\delta}{2} \qquad (5.7)$$

on the initial excess $E(Du; x_0, R)$ is satisfied, then the right-hand side of the previous inequality is estimated by $\delta \sup_{B_R(x_0)} |D\varphi|$ from above, and hence, we may apply the $\mathcal{A}$-harmonic approximation Lemma 4.27 (with $\gamma = 0$). Consequently, we find a function $h \in W^{1,2}(B_R(x_0), \mathbb{R}^N)$ which is $\mathcal{A}$-harmonic in $B_R(x_0)$ and which satisfies

$$R^{-n-2} \int_{B_R(x_0)} |w - h|^2 \, dx \leq \varepsilon \qquad \text{and} \qquad R^{-n} \int_{B_R(x_0)} |Dh|^2 \, dx \leq 1. \quad (5.8)$$

Since h solves a linear elliptic system with constant coefficients, the results from the linear theory apply. In particular, we know that h is locally smooth, and moreover, all excess decay estimates from Lemma 4.11 are available. We here note that we are interested in (Hölder-)continuity of the gradient, and not of the solution itself as before in Sect. 4.3.2. For this reason we take advantage of the decay estimates for Dh (and not for h). In this situation, Lemma 4.11 yields for every $r \in (0, R]$ the following two fundamental estimates:

$$\fint_{B_r(x_0)} |Dh|^2 \, dx \leq c \left(\frac{r}{R}\right)^n \fint_{B_R(x_0)} |Dh|^2 \, dx, \quad (5.9)$$

$$\fint_{B_r(x_0)} |Dh - (Dh)_{B_r(x_0)}|^2 \, dx \leq c \left(\frac{r}{R}\right)^{n+2} \fint_{B_R(x_0)} |Dh|^2 \, dx, \quad (5.10)$$

with a constant c depending only on n, N, and L, but we emphasize that rather γh than h is the relevant function, due to the rescaling of u. Next we want to carry these decay estimates over to the gradient of the weak solution u (or its rescaled version w). However, a priori, the $\mathcal{A}$-harmonic approximation lemma allows only to compare the L^2-distance of h and w, and not of their gradients. Nevertheless, with the Caccioppoli inequality at hand, this estimate on the level of the function u (or w) and not the gradient turns out to be sufficient to find a decay estimate for the excess of Du. For this purpose, we now consider $\tau \in (0, 1/2)$. In view of the minimizing property of the mean value for the map $\zeta \mapsto \int_\Omega |\varphi - \zeta|^2 \, dx$ for every $\varphi \in L^2(\Omega, \mathbb{R}^N)$, the Caccioppoli inequality from Proposition 5.4 allows to infer

$$E(Du; x_0, \tau R) = \fint_{B_{\tau R}(x_0)} |Du - (Du)_{B_{\tau R}(x_0)}|^2 \, dx$$

$$\leq \fint_{B_{\tau R}(x_0)} |Du - z_0|^2 \, dx$$

$$\leq c(n, L)(2\tau R)^{-2} \fint_{B_{2\tau R}(x_0)} |u - (u)_{B_{2\tau R}(x_0)} - z_0(x - x_0)|^2 \, dx$$

$$+ c(n, L)(1 + |z_0|)^{\frac{2}{1-\alpha}} (2\tau R)^{2\alpha} \quad (5.11)$$

for each $z_0 \in \mathbb{R}^{Nn}$, and we next need to make a good choice for z_0. Using once again the minimizing property of the mean value as above, the definition of w, and Poincaré's inequality, we compute

$$\fint_{B_{2\tau R}(x_0)} |u - (u)_{B_{2\tau R}(x_0)} - ((Du)_{B_R(x_0)} + \gamma(Dh)_{B_{2\tau R}(x_0)})(x - x_0)|^2 \, dx$$

$$\leq \fint_{B_{2\tau R}(x_0)} |u - \gamma(h)_{B_{2\tau R}(x_0)} - ((Du)_{B_R(x_0)} + \gamma(Dh)_{B_{2\tau R}(x_0)})(x - x_0)|^2 \, dx$$

$$= \gamma^2 \fint_{B_{2\tau R}(x_0)} |w - (h)_{B_{2\tau R}(x_0)} - (Dh)_{B_{2\tau R}(x_0)}(x - x_0)|^2 \, dx$$

$$\leq 2\gamma^2 \Big[\fint_{B_{2\tau R}(x_0)} |w - h|^2 \, dx$$

$$+ \fint_{B_{2\tau R}(x_0)} |h - (h)_{B_{2\tau R}(x_0)} - (Dh)_{B_{2\tau R}(x_0)}(x - x_0)|^2 \, dx \Big]$$

$$\leq c(n, N)\gamma^2 \Big[(2\tau R)^{-n} \int_{B_R(x_0)} |w - h|^2 \, dx$$

$$+ (2\tau R)^{2-n} \int_{B_{2\tau R}(x_0)} |Dh - (Dh)_{B_{2\tau R}(x_0)}|^2 \, dx \Big] .$$

With the decay estimate for Dh from (5.10) as well as with both the ε-closeness of w and h in the $L^2(B_R(x_0), \mathbb{R}^N)$-sense and the $L^2(B_R(x_0), \mathbb{R}^{Nn})$ bound on Dh in (5.8), we then find

$$\fint_{B_{2\tau R}(x_0)} |u - (u)_{B_{2\tau R}(x_0)} - ((Du)_{B_R(x_0)} + \gamma(Dh)_{B_{2\tau R}(x_0)})(x - x_0)|^2 \, dx$$

$$\leq c(n, N, L)\gamma^2 \Big[\tau^{-n} R^2 \varepsilon + \tau^4 R^{2-n} \int_{B_R(x_0)} |Dh|^2 \, dx \Big]$$

$$\leq c(n, N, L)(2\tau R)^2 \gamma^2 \big[\tau^{-n-2} \varepsilon + \tau^2 \big] .$$

In combination with (5.11) for the choice $z_0 = (Du)_{B_R(x_0)} + \gamma(Dh)_{B_{2\tau R}(x_0)}$, this gives

$$E(Du; x_0, \tau R) \leq c(n, N, L)\gamma^2 \big[\tau^{-n-2}\varepsilon + \tau^2 \big]$$

$$+ c(n, L)(1 + |(Du)_{B_R(x_0)} + \gamma(Dh)_{B_{2\tau R}(x_0)}|)^{\frac{2}{1-\alpha}} (2\tau R)^{2\alpha} .$$

Now, employing the definition of γ from (5.6) and taking into account the boundedness of the mean values of Dh via (5.9), we can continue to estimate the excess of Du, and we arrive at

$$E(Du; x_0, \tau R) \leq c_*(n, N, L)\big[\tau^{-n-2}\varepsilon + \tau^2\big] E(Du; x_0, R)$$
$$+ c(n, N, L)\big[\tau^{-n-2}\varepsilon + \tau^2\big]\delta^{-2}(1 + |(Du)_{B_R(x_0)}|)^{2(1+\alpha)} R^{2\alpha}$$
$$+ c(n, N, L)(1 + |(Du)_{B_R(x_0)}| + \gamma)^{\frac{2}{1-\alpha}}(\tau R)^{2\alpha}. \qquad (5.12)$$

Now we fix the free parameters τ and ε as follows: we first choose $\tau \in (0, 1/2)$ such that

$$2c_*\tau^2 \leq \tau^{1+\alpha}$$

holds, where c_* is the constant from the previous inequality. This fixes τ in dependency of n, N, L, and α. Next, we set $\varepsilon = \tau^{n+4}$. This determines ε and in turn the parameter δ in dependency of n, N, L, and α. As a consequence of these choices and the assumed bound $|(Du)_{B_R(x_0)}| \leq 2M_0$, the previous estimate (5.12) and once again the definition of γ yield the desired preliminary decay estimate

$$E(Du; x_0, \tau R) \leq \tau^{1+\alpha} E(Du; x_0, R) + c(n, N, L, \alpha, M_0)(\tau R)^{2\alpha},$$

and the latter constant might blow up when $M_0 \nearrow \infty$ or when $\alpha \nearrow 1$. Since this decay estimate was proved to hold under the one and only smallness assumption (5.7), which can be rephrased as a smallness condition on the initial excess and the function $\tilde{\omega}$, the proof of the lemma is complete. □

Remark 5.7 In contrast to the excess decay estimate (4.17), which was achieved before via the three different comparison techniques (for the toy case of elliptic systems with vector fields that are linear in the gradient variable), not only the initial excess appears on the right-hand side of the excess decay estimate (5.5), but also an expression involving the initial mean value $(Du)_{B_R(x_0)}$. For this reason, in order to iterate this preliminary excess decay estimate, it is necessary to control the mean values of Du on balls of different scales, i.e., we need to bound $|(Du)_{B_{\tau^k R}(x_0)}|$ for all $k \in \mathbb{N}$.

An iteration argument, which relies on the previous Lemma (5.6), then yields the final excess decay estimate, provided that the initial excess is sufficiently small.

Lemma 5.8 *Let $M_0 > 0$. There exist constants $R_0 = R_0(n, N, L, \alpha, M_0) \in (0, 1)$ and $\varepsilon_0 = \varepsilon_0(n, N, L, \alpha, \tilde{\omega}, M_0) \in (0, 1)$ such that the following statement is true: whenever $u \in W^{1,2}(\Omega, \mathbb{R}^N)$ is a weak solution to the system (5.1) under the assumptions (H1), (H2), (H3), and (H4) such that for some ball $B_R(x_0) \subset \Omega$ with $R \leq R_0$ we have*

$$E(Du; x_0, R) \leq \varepsilon_0^2 \qquad and \qquad |(Du)_{B_R(x_0)}| \leq M_0, \qquad (5.13)$$

then, for all $r \in (0, R]$, we have the decay estimate

$$E(Du; x_0, r) \le c(n, N, L, \alpha, M_0)\left(\left(\frac{r}{R}\right)^{2\alpha} E(Du; x_0, R) + r^{2\alpha}\right). \quad (5.14)$$

Proof The strategy of proof for such a result – also in more general situations – is based on an iteration of a preliminary excess decay estimate, which is here provided by Lemma 5.6. Such an iteration argument requires in general that the assumptions of the preliminary excess decay estimate are satisfied in each iteration step, which here means that we need to verify on the one hand that the excesses of Du satisfy the same smallness condition, and on the other hand that the mean values of Du remain uniformly bounded. The first requirement follows immediately from the preliminary excess decay estimate and a suitable choice of the maximal radius R_0, whereas the second requirement is achieved by taking advantage of the relation between mean values on two different balls and the excess on the bigger ball. In this way, we obtain the excess decay estimate in (5.14) for every value r of the form $\tau^k R$, for a fixed number $\tau \in (0, 1)$ (stemming from the preliminary excess decay estimate) and all $k \in \mathbb{N}$. The final estimate for an arbitrary radius $r \in (0, R]$ then follows by a standard interpolation argument.

We start by fixing the two parameters $\tau = \tau(n, N, L, \alpha) \in (0, 1)$ and $\tilde{\varepsilon}_0 = \tilde{\varepsilon}_0(n, N, L, \alpha, \tilde{\omega}) \in (0, 1)$ according to Lemma 5.6. We now select first $\varepsilon_0 \in (0, 1]$ and then $R_0 \in (0, 1]$ such that

$$\varepsilon_0 \le \min\left\{\tilde{\varepsilon}_0, M_0(1 - \tau^\alpha)\tau^{n/2}\right\} \quad \text{and} \quad c_{Dec} R_0^{2\alpha} \le (1 - \tau^{1-\alpha})\varepsilon_0^2 \quad (5.15)$$

are satisfied, where $c_{Dec} = c_{Dec}(n, N, L, \alpha, M_0)$ denotes the constant from the estimate (5.5) (with M_0 fixed as in the hypotheses of Lemma 5.8). With these smallness conditions in mind, we now check by induction that the estimates

$$E(Du; x_0, \tau^k R) \le \tau^{(1+\alpha)k} E(Du; x_0, R) + c_{Dec}\frac{1 - \tau^{(1-\alpha)k}}{1 - \tau^{1-\alpha}}(\tau^k R)^{2\alpha}, \quad (5.16)$$

$$E(Du; x_0, \tau^k R) \le \tau^{2\alpha k}\varepsilon_0^2 \quad \text{and} \quad |(Du)_{B_{\tau^k R}(x_0)}| \le 2M_0 \quad (5.17)$$

are true for all $k \in \mathbb{N}_0$, provided that the initial conditions in (5.13) are fulfilled for the initial ball $B_R(x_0) \subset \Omega$ with $R \le R_0$.

Obviously, $(5.16)_0$ always holds, while the two bounds in $(5.17)_0$ are already satisfied by the assumptions in (5.13). Let us now suppose that $(5.16)_j$ and $(5.17)_j$ are true for all $j \in \{0, \ldots, k-1\}$ and some $k \in \mathbb{N}$, and we need to verify $(5.16)_k$ and $(5.17)_k$. To this end, we first observe that, for every $j \in \{0, \ldots, k-1\}$, we have the preliminary decay estimate (5.5) for R replaced by $\tau^j R$ at our disposal, as a consequence of $(5.17)_j$. Combining this decay estimate for $j = k - 1$ with $(5.16)_{k-1}$ immediately gives $(5.16)_k$:

$$E(Du; x_0, \tau^k R) \leq \tau^{1+\alpha} E(Du; x_0, \tau^{k-1} R) + c_{Dec}(\tau^k R)^{2\alpha}$$

$$\leq \tau^{(1+\alpha)k} E(Du; x_0, R)$$

$$+ \tau^{1+\alpha} c_{Dec} \frac{1 - \tau^{(1-\alpha)(k-1)}}{1 - \tau^{1-\alpha}}(\tau^{k-1}R)^{2\alpha} + c_{Dec}(\tau^k R)^{2\alpha}$$

$$= \tau^{(1+\alpha)k} E(Du; x_0, R) + c_{Dec}\frac{1 - \tau^{(1-\alpha)k}}{1 - \tau^{1-\alpha}}(\tau^k R)^{2\alpha}.$$

We next check the conditions in $(5.17)_k$. Employing $E(Du; x_0, R) \leq \varepsilon_0^2$ and the smallness condition on R_0 from (5.15) in the latter inequality, we deduce the first assertion in $(5.17)_k$:

$$E(Du; x_0, \tau^k R) \leq \tau^{(1+\alpha)k} E(Du; x_0, R) + c_{Dec}\frac{1 - \tau^{(1-\alpha)k}}{1 - \tau^{1-\alpha}}\tau^{2\alpha k} R_0^{2\alpha}$$

$$\leq \tau^{(1+\alpha)k}\varepsilon_0^2 + \left(1 - \tau^{(1-\alpha)k}\right)\tau^{2\alpha k}\varepsilon_0^2 = \tau^{2\alpha k}\varepsilon_0^2.$$

In order to prove the second assertion in $(5.17)_k$, we use Jensen's inequality, the initial bound $|(Du)_{B_R(x_0)}| \leq M_0$ from (5.14), the first bound in $(5.17)_j$ for every $j \in \{0, \ldots, k-1\}$ and the smallness condition on ε_0 in (5.15), and we find

$$|(Du)_{B_{\tau^k R}(x_0)}| \leq |(Du)_{B_R(x_0)}| + \sum_{j=0}^{k-1} |(Du)_{B_{\tau^j R}(x_0)} - (Du)_{B_{\tau^{j+1} R}(x_0)}|$$

$$\leq M_0 + \tau^{-\frac{n}{2}} \sum_{j=0}^{k-1} \left(E(Du; x_0, \tau^j R)\right)^{\frac{1}{2}}$$

$$\leq M_0 + \tau^{-\frac{n}{2}} \sum_{j=0}^{\infty} \tau^{\alpha j}\varepsilon_0 \leq 2M_0.$$

To complete the proof of the lemma, it still remains to establish a continuous version of the excess decay estimate (5.16). To this end, we consider an arbitrary radius $r \in (0, R]$ and determine the unique number $k \in \mathbb{N}_0$ such that $\tau^{k+1} R < r \leq \tau^k R$. With $(5.16)_k$ we then infer

$$E(Du; x_0, r) \leq \left(\frac{\tau^k R}{r}\right)^n E(Du; x_0, \tau^k R)$$

$$\leq \left(\frac{\tau^k R}{r}\right)^n \left(\tau^{2\alpha k} E(Du; x_0, R) + c_{Dec}\frac{1 - \tau^{(1-\alpha)k}}{1 - \tau^{1-\alpha}}(\tau^k R)^{2\alpha}\right)$$

$$\leq \tau^{-n-2\alpha}\left(\left(\frac{r}{R}\right)^{2\alpha} E(Du; x_0, R) + c_{Dec}\frac{1 - \tau^{(1-\alpha)k}}{1 - \tau^{1-\alpha}}r^{2\alpha}\right).$$

This yields the assertion (5.14), when we take into account that the constants c_{Dec} and τ depend only on the parameters n, N, L, α, and M_0.   □

The previous lemma states that an excess decay estimate for the gradient of a weak solution u holds, whenever, in the limit $R \searrow 0$, on the one hand the mean value of Du on $B_R(x_0)$ remains bounded and on the other hand the excess on $B_R(x_0)$ can be made arbitrarily small. These requirements now allow us to characterize the set of singular points of Du and to state the announced partial regularity result for Du, which is the main result of this section. This statement was obtained first by Giaquinta and Modica [36] via the direct method (at the same time, Ivert [48] proved in a similar way partial regularity for bounded solutions to inhomogeneous elliptic systems). Later on, alternative proofs were given by Hamburger [45] via a blow-up technique, and by Duzaar and Grotowski [21] via the method of $\mathcal{A}$-harmonic approximation, which we have presented here. The latter approach has the advantage that the optimal $C^{1,\alpha}$-regularity of weak solutions (see Example 5.10 for the optimality of the Hölder exponent) follows in only one step, while in the original proof [36] initially $C^{1,\delta}$-regularity is proved (for some $\delta > 0$ resulting from the higher integrability of the gradients), which only in a second step is improved to the optimal Hölder exponent α.

Theorem 5.9 (Giaquinta and Modica; Ivert)   *Let $u \in W^{1,2}(\Omega, \mathbb{R}^N)$ be a weak solution to the system (5.1) with a vector field $a \colon \Omega \times \mathbb{R}^N \times \mathbb{R}^{Nn} \to \mathbb{R}^{Nn}$ satisfying the assumptions (H1), (H2), (H3), and (H4). Then we have the characterization of the singular set $\mathrm{Sing}_0(Du) = \Sigma_1 \cup \Sigma_2$ via*

$$\Sigma_1 := \left\{ x_0 \in \Omega \colon \liminf_{\varrho \searrow 0} \fint_{\Omega(x_0, \varrho)} |Du - (Du)_{\Omega(x_0, \varrho)}|^2 \, dx > 0 \right\},$$

$$\Sigma_2 := \left\{ x_0 \in \Omega \colon \limsup_{\varrho \searrow 0} |(Du)_{\Omega(x_0, \varrho)}| = \infty \right\},$$

and in particular, $\mathcal{L}^n(\mathrm{Sing}_0(Du)) = 0$. Moreover, we have $\mathrm{Reg}_0(Du) = \mathrm{Reg}_\beta(Du)$ for every $\beta \in (0, \alpha]$, i.e. $u \in C^{1,\alpha}(\mathrm{Reg}_0(Du), \mathbb{R}^N)$.

Proof Given $x_0 \in \Omega \setminus (\Sigma_1 \cup \Sigma_2)$, we can find $M_0 > 0$ and ball $B_R(x_0) \Subset \Omega$ with $R < R_0(n, N, L, \alpha, M_0)$ such that

$$E(Du; x_0, R) < \varepsilon_0^2 \qquad \text{and} \qquad |(Du)_{B_R(x_0)}| < M_0$$

hold, for R_0 and ε_0 the constants chosen according to Lemma 5.8. Moreover, since $B_R(x_0)$ is compactly contained in Ω and since the functions $y \mapsto E(Du; y, R)$ and $y \mapsto (Du)_{B_R(y)}$ are continuous for every fixed $R > 0$, we find a small neighbourhood $B_\delta(x_0) \subset B_R(x_0)$ of x_0 such that $B_R(y) \subset \Omega$ holds and such that the bounds

$$E(Du; y, R) < \varepsilon_0^2 \qquad \text{and} \qquad |(Du)_{B_R(y)}| < M_0$$

are satisfied for all $y \in B_\delta(x_0)$. Hence, the application of the previous Lemma 5.8 yields

$$E(Du; y, r) \leq c(n, N, L, \alpha, M_0)\left(\left(\frac{r}{R}\right)^{2\alpha} E(Du; y, R) + r^{2\alpha}\right)$$

for all $y \in B_\delta(x_0)$, and consequently, local $C^{1,\alpha}$-regularity of Du in $B_\delta(x_0)$ follows from Theorem 1.27 with $p = 2$ (and in combination with Corollary 1.63). This proves $\Omega \setminus (\Sigma_1 \cup \Sigma_2) \subset \text{Reg}_\alpha(Du) \subset \text{Reg}_0(Du)$, and the claimed characterization then follows from the obvious inclusion $\text{Reg}_0(Du) \subset \Omega \setminus (\Sigma_1 \cup \Sigma_2)$. $\qquad\square$

We finally observe that the Hölder regularity exponent α, which was found for Du in the last Theorem 5.9, is the one given by the regularity of the vector field a with respect to the first and second variable. In fact, this is the optimal result: the regularity of Du cannot be expected to be better than the one of the vector field, as the following example shows. In particular, we emphasize that this fact is true already in the scalar case (and moreover, the vector field given below is linear in the gradient variable and does not depend explicitly on the solution).

Example 5.10 ([43], Example 1.1) *Let* $n \geq 2$, $N = 1$ *and* $\alpha \in (0, 1)$. *The vector field* $a: B_1 \times \mathbb{R}^n \to \mathbb{R}^n$ *defined via*

$$a(x, z) := \frac{z}{1 + x_n^\alpha}.$$

satisfies the assumptions (H1), (H2), (H3), *and* (H4), *and the function*

$$u(x) = x_n^{1+\alpha} + (1 + \alpha)x_n$$

is a weak solution to $\text{div}\, a(x, Du) = 0$ *in* B_1. *Furthermore,* u *is of class* $C^{1,\alpha}(B_1)$, *but no more regular.*

For comparison: a partial regularity result for minimizers to convex variational integrals Concerning partial C^1-regularity results, we would also like to address very briefly a related result for minimizers of variational integrals of the form

$$F[w; \Omega] := \int_\Omega f(x, w, Dw)\, dx. \tag{5.18}$$

We here suppose on the integrand f suitable growth and continuity assumptions, and most importantly, strict convexity in the gradient variable, see further below. For simplicity we here restrict ourselves to the case of quadratic growth in the gradient variable (which then is a straightforward extension of the case of quadratic variational integrals mentioned in Chap. 4.3.3), and we wish to investigate the regularity of minimizers in Dirichlet classes of $W^{1,2}(\Omega, \mathbb{R}^N)$. Again, we want to work under low regularity assumptions on the integrand, which in general do not permit to take advantage of the associated Euler–Lagrange system. Therefore, we need to develop the regularity proof with new ingredients, which rather rely on the minimization property and not on the Euler–Lagrange system, but which nevertheless shares the same intermediate steps as in the case of quasilinear elliptic systems. We now implement a simple version of the method of $\mathcal{A}$-harmonic version given by Schmidt [74], under the following assumptions:

(F0) f is coercive and satisfies a quadratic growth condition

$$|z|^2 \leq f(x, u, z) \leq L(1 + |z|)^2 ,$$

(F1) f is of class C^2 with respect to z with

$$|D_z^2 f(x, u, z)| \leq L ,$$

(F2) f is strictly convex in z with

$$D_z^2 f(x, u, z)\xi \cdot \xi \geq |\xi|^2 \qquad \text{for all } \xi \in \mathbb{R}^{Nn} ,$$

(F3) f and $D_z f$ are Hölder continuous in u and (x, u) with

$$|f(x, u, z) - f(x, \tilde{u}, z)| \leq 2L\omega_{\alpha_1}\big(|u - \tilde{u}|\big)(1 + |z|)^2 ,$$

$$|D_z f(x, u, z) - D_z f(\tilde{x}, \tilde{u}, z)| \leq 2L\omega_{\alpha_2}\big(|x - \tilde{x}| + |u - \tilde{u}|\big)(1 + |z|) ,$$

(F4) $D_z^2 f$ is continuous with

$$|D_z^2 f(x, u, z) - D_z^2 f(x, u, \tilde{z})| \leq 2L\tilde{\omega}\big(|z - \tilde{z}|\big) ,$$

for all $x, \tilde{x} \in \Omega$, $u, \tilde{u} \in \mathbb{R}^N$ and $z, \tilde{z} \in \mathbb{R}^{Nn}$, with a constant $L \geq 1$. Here ω_{α_1}, ω_{α_2} and $\tilde{\omega}$ are moduli of continuity, bounded by 1 from above, monotonically non-decreasing and concave. Furthermore, we suppose $\omega_{\alpha_i}(t) \leq \min\{1, t^{\alpha_i}\}$ for $i \in \{1, 2\}$ and exponents $\alpha_1, \alpha_2 \in (0, 1)$ and $\lim_{t \searrow 0} \tilde{\omega}(t) = \tilde{\omega}(0) = 0$.

Remarks 5.11

(i) Under the assumptions (F0) and (F2) one automatically has a linear growth condition on $D_z f$ with respect to the gradient variable. This ensures in particular that all integrals appearing later on are well-defined.

(ii) The Hölder continuity assumption (F3) is more general than the requirement that f is Hölder continuous in (x, u) with

$$|f(x, u, z) - f(\tilde{x}, \tilde{u}, z)| \leq 2L\omega_\alpha\big(|x - \tilde{x}| + |u - \tilde{u}|\big)(1 + |z|), \qquad (5.19)$$

provided that in addition (F1) is supposed. In fact, in this case the Hölder continuity condition on $D_z f$ can be shown to be true with exponent $\alpha_2 = \alpha/2$ (possibly after choosing L larger), see [74, Appendix A].

Under the assumptions (F0), (F1), (F2), (F3), and (F4), partial C^1-regularity of minimizers to the functional F in Dirichlet classes can be shown. This was first achieved by Giaquinta and Giusti [33] via a version of the direct method (with a similar reasoning as in Sect. 4.3.3), but we here present the proof via the method of $\mathcal{A}$-harmonic approximation, as suggested by Schmidt [74]. As before in the situation of elliptic systems, this approach has the advantage that the optimal Hölder exponent is achieved in only one step. However, we emphasize that the partial regularity result holds in fact under the weaker assumption of strict quasiconvexity instead of convexity, as it was shown by Giaquinta and Modica [38] via a the direct method and, simultaneously, by Fusco and Hutchinson [27] via the blow-up technique.

Theorem 5.12 (Giaquinta and Giusti)   *Let $u \in W^{1,2}(\Omega, \mathbb{R}^N)$ be a minimizer of the functional F with a Carathéodory integrand f satisfying* (F0), (F1), (F2), (F3), *and* (F4). *Then we have the characterization of the singular set* $\mathrm{Sing}_0(Du) = \Sigma_1 \cup \Sigma_2$ *via*

$$\Sigma_1 := \Big\{ x_0 \in \Omega \colon \liminf_{\varrho \searrow 0} \fint_{\Omega(x_0,\varrho)} |Du - (Du)_{\Omega(x_0,\varrho)}|^2 \, dx > 0 \Big\},$$

$$\Sigma_2 := \Big\{ x_0 \in \Omega \colon \limsup_{\varrho \searrow 0} |(Du)_{\Omega(x_0,\varrho)}| = \infty \Big\},$$

and in particular, $\mathcal{L}^n(\mathrm{Sing}_0(Du)) = 0$. *Moreover, we have* $\mathrm{Reg}_0(Du) = \mathrm{Reg}_\beta(Du)$ *for every* $\beta \in (0, \tilde{\alpha}]$ *with*

$$\tilde{\alpha} := \min\{\alpha_1/(2-\alpha_1), \alpha_2\}, \tag{5.20}$$

and therefore $u \in C^{1,\tilde{\alpha}}(\mathrm{Reg}_0(Du), \mathbb{R}^N)$.

The Hölder exponent (5.20) appeared for the first time in a paper by Phillips [73] and is sharp. Moreover, in the case of the sole Hölder continuity condition (5.19) on f instead of (F3) (which is precisely the setting of the papers [27, 33, 38]), we end up with partial $C^{1,\alpha/2}$-regularity of minimizers.

Similarly as for the elliptic systems considered before, the starting point for the proof of Theorem 5.12 is a Caccioppoli inequality for affine perturbations of the minimizer.

Proposition 5.13 (Caccioppoli inequality)   *Let $u \in W^{1,2}(\Omega, \mathbb{R}^N)$ be a minimizer of the functional F with with a Carathéodory integrand f satisfying* (F0), (F1), (F2), *and* (F3). *Then, for all* $\zeta \in \mathbb{R}^N$, $z_0 \in \mathbb{R}^{Nn}$

and $B_r(x_0) \subset \Omega$ with $r \le 1$, we have

$$\int_{B_{r/2}(x_0)} |Du - z_0|^2 \, dx \le c(L) r^{-2} \int_{B_r(x_0)} |u - \zeta - z_0(x - x_0)|^2 \, dx$$

$$+ c(L)(1 + |z_0|)^{\frac{4}{1 - \alpha_2}} r^{n + 2\tilde{\alpha}} .$$

Proof We consider $r/2 \le \varrho < \sigma \le r$ and take a standard cut-off function $\eta \in C_0^\infty(B_\sigma(x_0), [0, 1])$ which satisfies $\eta \equiv 1$ in $B_\varrho(x_0)$ and $|D\eta| \le 2(\sigma - \varrho)^{-1}$. We now define

$$\varphi := \eta(u - \zeta - z_0(x - x_0)) \quad \text{and} \quad \psi := (1 - \eta)(u - \zeta - z_0(x - x_0)),$$

and we note $\varphi \in W_0^{1,2}(B_\sigma(x_0), \mathbb{R}^N)$, $\psi \equiv 0$ in $B_\varrho(x_0)$ and

$$D\varphi + D\psi = Du - z_0 . \tag{5.21}$$

We now want to test the minimality of u against the competitor $u - \varphi$. To this end we first use the convexity assumption (F2) to estimate

$$\int_{B_\sigma(x_0)} |D\varphi|^2 \, dx$$

$$\le \int_{B_\sigma(x_0)} \int_0^1 \int_0^1 D_z^2 f(x, u, z_0 + st D\varphi) \, ds \, dt D\varphi \cdot D\varphi \, dx$$

$$= \int_{B_\sigma(x_0)} \left[f(x, u, z_0 + D\varphi) - f(x, u, z_0) - D_z f(x, u, z_0) \cdot D\varphi \right] dx$$

$$= \int_{B_\sigma(x_0)} \left[f(x, u, z_0 + D\varphi) - f(x, u, Du) \right] dx$$

$$+ \int_{B_\sigma(x_0)} \left[f(x, u, Du) - f(x, u - \varphi, Du - D\varphi) \right] dx$$

$$+ \int_{B_\sigma(x_0)} \left[f(x, u - \varphi, Du - D\varphi) - f(x, u - \varphi, z_0) \right] dx$$

$$+ \int_{B_\sigma(x_0)} \left[f(x, u - \varphi, z_0) - f(x, u, z_0) \right] dx$$

$$+ \int_{B_\sigma(x_0)} \left[D_z f(x_0, \zeta, z_0) - D_z f(x, u, z_0) \right] \cdot D\varphi \, dx$$

$$=: I + II + III + IV + V .$$

We first observe that the minimality of u guarantees

$$II \leq 0.$$

Next, we consider the terms I and III, which are actually integrals on the annulus $B_\sigma(x_0) \setminus B_\varrho(x_0)$. This is easily seen by rewriting $z_0 + D\varphi$ in the first integral as $Du - D\psi$ and by rewriting $Du - D\varphi$ in the second integral as $z_0 + D\psi$, respectively, and by keeping in mind that $\psi \equiv 0$ holds in $B_\varrho(x_0)$. In order to obtain a bound in terms of the affine perturbation of u (or its gradient) we then rewrite the sum $I + III$ as follows

$$I + III$$

$$= \int_{B_\sigma(x_0)\setminus B_\varrho(x_0)} \int_0^1 \Big[- D_z f(x, u, Du - tD\psi)$$

$$+ D_z f(x, u - \varphi, z_0 + tD\psi) \Big] \, dt \cdot D\psi \, dx$$

$$= - \int_{B_\sigma(x_0)\setminus B_\varrho(x_0)} \int_0^1 \int_0^1 D_z^2 f(x, u, z_0 + s(Du - z_0 - tD\psi)) \, ds$$

$$\times (Du - z_0 - tD\psi) \, dt \cdot D\psi \, dx$$

$$+ \int_{B_\sigma(x_0)\setminus B_\varrho(x_0)} \int_0^1 \int_0^1 D_z^2 f(x, u - \varphi, z_0 + stD\psi) \, ds \, t \, dt D\psi \cdot D\psi \, dx$$

$$+ \int_{B_\sigma(x_0)\setminus B_\varrho(x_0)} \big[D_z f(x, u - \varphi, z_0) - D_z f(x, u, z_0) \big] \cdot D\psi \, dx.$$

At this stage we take advantage of the boundedness of $D_z^2 f$ according to assumption (F1) and of the Hölder continuity of $D_z f$ according to assumption (F3), which allow us to infer the bound

$$I + III \leq c(L) \int_{B_\sigma(x_0)\setminus B_\varrho(x_0)} \big[|Du - z_0|^2 + |D\psi|^2 \big] \, dx$$

$$+ c(L)(1 + |z_0|) \int_{B_\sigma(x_0)\setminus B_\varrho(x_0)} \omega_{\alpha_2}(|\varphi|) |D\psi| \, dx.$$

For the remaining integrals IV and V we again use the Hölder continuity conditions in (F3) and find

$$IV + V \leq c(L)(1 + |z_0|)^2 \int_{B_\sigma(x_0)} \omega_{\alpha_1}(|\varphi|) \, dx$$

$$+ c(L)(1 + |z_0|) \int_{B_\sigma(x_0)} \omega_{\alpha_2}(|x - x_0| + |u - \zeta|) |D\varphi| \, dx.$$

Collecting all estimates, using Young's inequality and recalling the formula (5.21), we obtain in a first step

$$
\int_{B_\sigma(x_0)} |D\varphi|^2 \, dx
$$

$$
\leq c(L) \int_{B_\sigma(x_0)\backslash B_\varrho(x_0)} \left[|Du - z_0|^2 + |D\psi|^2 \right] dx
$$

$$
+ c(L)(1 + |z_0|)^2 \int_{B_\sigma(x_0)} \omega_{\alpha_1}(|u - \zeta - z_0(x - x_0)|) \, dx
$$

$$
+ c(L)(1 + |z_0|)^2 \int_{B_\sigma(x_0)} \omega_{\alpha_2}^2\big((1 + |z_0|)r + |u - \zeta - z_0(x - x_0)|\big) \, dx .
$$

Then, keeping in mind the definitions of φ, ψ and the properties of η, we get in a second step

$$
\int_{B_\varrho(x_0)} |Du - z_0|^2 \, dx \leq c(L)\bigg[(\sigma - \varrho)^{-2} \int_{B_\sigma(x_0)} |u - \zeta - z_0(x - x_0)|^2 \, dx
$$

$$
+ (1 + |z_0|)^{\frac{4}{1 - \alpha_2}} r^{n + 2\tilde{\alpha}} + \int_{B_\sigma(x_0)\backslash B_\varrho(x_0)} |Du - z_0|^2 \, dx \bigg] ,
$$

$$
\tag{5.22}
$$

where, for obtaining the second term on the right-hand side, we have employed $\sigma - \varrho \leq r \leq 1$, Young's inequality and the estimate

$$
\min\left\{ (1 + |z_0|)^{\frac{4}{2 - \alpha_1}} r^{\frac{2\alpha_1}{2 - \alpha_1}}, (1 + |z_0|)^{2 + 2\alpha_2} r^{2\alpha_2}, (1 + |z_0|)^{\frac{2}{1 - \alpha_2}} r^{\frac{2\alpha_2}{1 - \alpha_2}} \right\}
$$

$$
\leq (1 + |z_0|)^{\frac{4}{1 - \alpha_2}} r^{2\tilde{\alpha}}
$$

with $\tilde{\alpha} = \min\{\alpha_1/(2 - \alpha_1), \alpha_2\}$ as defined in (5.20). Now we are again in a situation for the hole-filling argument: we add $c(L) \int_{B_\varrho(x_0)} |Du - z_0|^2 \, dx$ to both sides of the inequality (5.22), which then allows us to apply the iteration Lemma B.1, and the assertion then follows. $\square$

Lemma 5.14 (Approximate $\mathcal{A}$-harmonicity III) *Let $u \in W^{1,2}(\Omega, \mathbb{R}^N)$ be a minimizer of the functional F with a Carathéodory integrand f satisfying (F0), (F1), (F2), (F3), and (F4). Then, for all $z_0 \in \mathbb{R}^{Nn}$ and* *every $B_\varrho(x_0) \subset \Omega$ with $\varrho \leq 1$, we have*

$$\left| \varrho^{-n} \int_{B_\varrho(x_0)} D_z^2 f(x_0, (u)_{B_\varrho(x_0)}, z_0)(Du - z_0) \cdot D\varphi \, dx \right|$$

$$\leq c_{Ap}(n, N, L) \left[\tilde{\omega}^{1/2} \left(\left(\fint_{B_\varrho(x_0)} |Du - z_0|^2 \, dx \right)^{\frac{1}{2}} \right) \left(\fint_{B_\varrho(x_0)} |Du - z_0|^2 \, dx \right)^{\frac{1}{2}} \right.$$

$$\left. + \fint_{B_\varrho(x_0)} |Du - z_0|^2 \, dx + (1 + |z_0|)^2 \varrho^{\tilde{\alpha}} \right] \sup_{B_\varrho(x_0)} |D\varphi|$$

for all $\varphi \in C_0^1(B_\varrho(x_0), \mathbb{R}^N)$.

Proof We consider an arbitrary function $\varphi \in C_0^1(B_\varrho(x_0), \mathbb{R}^N)$ which satisfies, without loss of generality, $\sup_{B_\varrho(x_0)} |D\varphi| \leq 1$. With

$$\fint_{B_\varrho(x_0)} D_z f(x_0, (u)_{B_\varrho(x_0)}, z_0) \cdot D\varphi \, dx = 0$$

we first notice, for a parameter $\varsigma \in (0, 1]$ to be chosen later, the identity

$$\fint_{B_\varrho(x_0)} D_z^2 f(x_0, (u)_{B_\varrho(x_0)}, z_0)(Du - z_0) \cdot D\varphi \, dx$$

$$= \fint_{B_\varrho(x_0)} \int_0^1 \left[D_z^2 f(x_0, (u)_{B_\varrho(x_0)}, z_0) \right.$$

$$\left. - D_z^2 f(x_0, (u)_{B_\varrho(x_0)}, z_0 + t(Du - z_0)) \right] dt (Du - z_0) \cdot D\varphi \, dx$$

$$+ \fint_{B_\varrho(x_0)} \int_0^\varsigma \left[D_z f(x_0, (u)_{B_\varrho(x_0)}, Du) \right.$$

$$\left. - D_z f(x_0, (u)_{B_\varrho(x_0)}, Du - t D\varphi) \right] dt \cdot D\varphi \, dx$$

$$+ \fint_{B_\varrho(x_0)} \int_0^\varsigma D_z f(x_0, (u)_{B_\varrho(x_0)}, Du - t D\varphi) \, dt \cdot D\varphi \, dx$$

$$=: I + II + III.$$

With the help of Hölder's and Jensen's inequality, we deduce from (F4), exactly as in the proof of Lemma 5.5 on approximate $\mathcal{A}$-harmonicity for elliptic systems, the estimate

$$I \leq 2L\tilde{\omega}^{1/2} \left(\left(\fint_{B_\varrho(x_0)} |Du - z_0|^2 \, dx \right)^{\frac{1}{2}} \right) \left(\fint_{B_\varrho(x_0)} |Du - z_0|^2 \, dx \right)^{\frac{1}{2}}.$$

In view of assumption (F1), we get for the second term

$$II = \fint_{B_\varrho(x_0)} \fint_0^\varsigma \int_0^1 D_z^2 f(x_0, (u)_{B_\varrho(x_0)}, Du - stD\varphi)\, ds\, t\, dt D\varphi \cdot D\varphi\, dx \le L\varsigma\,.$$

Hence, it only remains to estimate the integral III, which we first rewrite as

$$III = \fint_{B_\varrho(x_0)} \fint_0^\varsigma \big[D_z f(x_0, (u)_{B_\varrho(x_0)}, Du - tD\varphi)$$
$$- D_z f(x, u, Du - tD\varphi)\big]\, dt \cdot D\varphi\, dx$$
$$+ \frac{1}{\varsigma} \fint_{B_\varrho(x_0)} \big[f(x, u, Du) - f(x, u - \varsigma\varphi, Du - \varsigma D\varphi)\big]\, dx$$
$$+ \frac{1}{\varsigma} \fint_{B_\varrho(x_0)} \big[f(x, u - \varsigma\varphi, Du - \varsigma D\varphi) - f(x, u, Du - \varsigma D\varphi)\big]\, dx\,.$$

Now, we employ the assumption (F3) on Hölder continuity of f and $D_z f$ to estimate the first and the third integral on the right-hand side, and we further use the minimality of u to see that the second integral is non-positive. This yields

$$III \le 2L \fint_{B_\varrho(x_0)} \omega_{\alpha_2}(|x - x_0| + |u - (u)_{B_\varrho(x_0)}|)(2 + |Du|)\, dx$$
$$+ \frac{2L}{\varsigma} \fint_{B_\varrho(x_0)} \omega_{\alpha_1}(|\varsigma\varphi|)(2 + |Du|)^2\, dx\,.$$

In order to continue to estimate the integrals on the right-hand side, we first observe that $\sup_{B_\varrho(x_0)} |\varphi| \le \varrho$ holds, because of $\varphi \in C_0^1(B_\varrho(x_0), \mathbb{R}^N)$ with $\sup_{B_\varrho(x_0)} |D\varphi| \le 1$. We further choose $\varsigma := \varrho^{\alpha_1/(2-\alpha_1)}$, which implies $\varsigma = \varsigma^{-1}(\varsigma\varrho)^{\alpha_1} \le 1$ (so that the upper bound of II and the second integral in the bound for III allow for the same scaling in ϱ). With these observations, combined with the application of the inequalities of Young and Poincaré, we then find

$$III \le c(L) \fint_{B_\varrho(x_0)} \big[|Du - z_0|^2 + \varrho^{-2}|u - (u)_{B_\varrho(x_0)}|^2\big]\, dx$$
$$+ c(L)\big[(1 + |z_0|)\varrho^{\alpha_2} + \varrho^{2\alpha_2} + (1 + |z_0|)^{\frac{2}{2-\alpha_2}} \varrho^{\frac{2\alpha_2}{2-\alpha_2}}$$
$$+ \varrho^{\frac{2\alpha_2}{1-\alpha_2}} + (1 + |z_0|)^2 \varrho^{\frac{\alpha_1}{2-\alpha_1}}\big]$$
$$\le c(n, N, L) \fint_{B_\varrho(x_0)} |Du - z_0|^2\, dx + c(L)(1 + |z_0|)^2 \varrho^{\tilde\alpha}\,.$$

Collecting the estimates for I, II, and III, we then arrive at the asserted upper bound, and the lower bound follows from the passage $\varphi \to -\varphi$. □

Now we have the main ingredients for the proof of partial C^1-regularity of minimizers at hand.

Sketch of proof of Theorem 5.12 Once the Caccioppoli inequality and the approximate $\mathcal{A}$-harmonicity lemma are established, the proof of partial regularity follows exactly along the line of arguments for elliptic systems (note that the estimates in Proposition 5.4 and Lemma 5.5 for weak solution to elliptic systems are the same as the ones in Proposition 5.13 and Lemma 5.14 obtained for minimizers of convex variational integrals). Therefore, we only comment on the necessary modifications. We start with a preliminary excess decay estimate as given in Lemma 5.6. For the application of the $\mathcal{A}$-harmonic approximation Lemma 4.27 we now define the bilinear form $\mathcal{A} \in R^{Nn \times Nn}$ and a rescaling parameter γ via

$$\mathcal{A} := D_z^2 f(x_0, (u)_{B_R(x_0)}, (Du)_{B_R(x_0)}),$$

$$\gamma := c_{Ap} \left[E(Du; x_0, R) + 4\delta^{-2}(1 + |(Du)_{B_R(x_0)}|)^4 R^{2\tilde\alpha} \right]^{\frac{1}{2}},$$

where $B_R(x_0) \subset \Omega$ is a ball with radius $R \leq 1$. This allows to show that the function $w(x) = \gamma^{-1}[u(x) - (Du)_{B_R(x_0)}(x - x_0)]$ is approximately $\mathcal{A}$-harmonic, provided that the smallness condition (5.7) on the initial excess $E(Du; x_0, R)$ is satisfied. From this stage we may proceed exactly as in the proof of Lemma 5.6, which results in the following statement: we find parameters $\tau = \tau(n, N, L, \alpha_1, \alpha_2) \in (0, 1)$ and $\tilde\varepsilon_0 = \tilde\varepsilon_0(n, N, L, \alpha_1, \alpha_2, \tilde\omega) \in (0, 1)$ such that

$$E(Du; x_0, \tau R) \leq \tau^{1+\tilde\alpha} E(Du; x_0, R) + c_{Dec}(n, N, L, \alpha, M_0)(\tau R)^{2\tilde\alpha}$$

holds, provided that we have

$$E(Du; x_0, R) \leq 2\tilde\varepsilon_0^2 \quad \text{and} \quad |(Du)_{B_R(x_0)}| \leq 2M_0.$$

This estimate can then be iterated as in Lemma 5.8 and results, for small initial excess and bounded initial mean value of Du, in the excess decay estimate (5.14), with exponent α replaced by $\tilde\alpha$. At this point, the characterization of the singular set $\mathrm{Sing}_0(Du)$ and the regularity improvement $\mathrm{Reg}_0(Du) = \mathrm{Reg}_{\tilde\alpha}(Du)$ follow as in the proof of Theorem 5.9. □

5.3   The Hausdorff Dimension of the Singular Set

So far we have proved two types of partial regularity results. On the one hand, in Sect. 4.3, we have obtained partial regularity of weak solutions u to particular quasilinear elliptic systems that are linear in the gradient variable, i.e., u is of class $C^{0,\alpha}$ on a set of full Lebesgue measure, for any $\alpha \in [0,1)$. Moreover, as a consequence of the characterization of regular points, the Hausdorff dimension of the singular set $\mathrm{Sing}_0(u)$ of u cannot exceed $n-2$ (while from the counterexamples presented in Sect. 4.1 it is clear that the Hausdorff dimensions of $\mathrm{Sing}_0(u)$ can in general not be less than $n-3$, see also Remark 4.4). On the other hand, in Sect. 5.2, we have studied general quasilinear elliptic systems in divergence form and have proved partial regularity of the gradient of weak solutions, i.e., u is of class $C^{1,\alpha}$ on an open set of full Lebesgue measure, and the exponent α from the Hölder continuity of the coefficients is the optimal one. Our next aim is to find also in this case estimates on the Hausdorff dimension of the singular sets. We are first going to show higher differentiability properties of weak solutions, which in turn will provide the bound $n-2\alpha$ for the Hausdorff dimension of the singular set $\mathrm{Sing}_0(Du)$ of Du. Then we discuss some Morrey type estimates that allow us to generalize the bound $n-2$ on the Hausdorff dimension for $\mathrm{Sing}_0(u)$, provided that the assumption $n \le 4$ of low dimensions is made.

5.3.1 Bounds in General Dimensions

We here continue to work under the permanent assumption of (H1), (H2), (H3), and (H4), and thus, the characterization of the set $\mathrm{Sing}_0(Du)$ of singular points of Du, which was obtained in Theorem 5.9, is available. The starting point for the discussion of the size of $\mathrm{Sing}_0(Du)$ is the following proposition concerning finite differences of Du, cf. [62, 63].

Proposition 5.15 (Preliminary estimate) *Let $u \in W^{1,2}(\Omega, \mathbb{R}^N)$ be a weak solution to the system (5.1) with a vector field $a\colon \Omega \times \mathbb{R}^N \times \mathbb{R}^{Nn} \to \mathbb{R}^{Nn}$ satisfying (H1), (H2), and (H3). Let $B_R(x_0) \subset \Omega$ be a ball, $s \in \{1,\dots,n\}$, and $h \in \mathbb{R}$ with $|h| < R/4$. Then the following statements are true:*

(i) *If a does not depend explicitly on the u-variable, i.e. $a(x,u,z) \equiv a(x,z)$ for all $x \in \Omega$, $u \in \mathbb{R}^N$, and $z \in \mathbb{R}^{Nn}$, then we have*

$$\int_{B_{R/2}(x_0)} |\tau_{s,h} Du(x)|^2 \, dx \le c(R,L)|h|^{2\alpha} \int_{B_R(x_0)} \left(1 + |Du(x)|^2\right) dx \,.$$

(ii) *In the general case we have*

$$\int_{B_{R/2}(x_0)} |\tau_{s,h} Du(x)|^2 \, dx \leq c(R,L)|h|^{2\alpha} \int_{B_R(x_0)} \left(1 + |Du(x)|^2\right) dx$$

$$+ c(R,L) \int_{B_{3R/4}(x_0)} |Du(x + he_s)|^2 \omega_\alpha^2(|\tau_{s,h} u(x)|) \, dx \, .$$

Proof We take a standard cut-off function $\eta \in C_0^\infty(B_{3R/4}(x_0), [0,1])$ satisfying $\eta \equiv 1$ in $B_{R/2}(x_0)$ and $|D\eta| \leq 8/R$. We now test the weak formulation of system (5.1) with the function $\varphi := \tau_{s,-h}(\eta^2 \tau_{s,h} u)$. Using the integration by parts formula for finite differences, we hence get

$$\int_\Omega \tau_{s,h} a(x, u(x), Du(x)) \cdot D(\eta^2(x)\tau_{s,h} u(x)) \, dx = 0 \, . \qquad (5.23)$$

Next we decompose the finite differences $\tau_{s,h} a(x, u(x), Du(x))$ into

$$\tau_{s,h} a(x, u(x), Du(x))$$
$$= a(x + he_s, u(x + he_s), Du(x + he_s)) - a(x, u(x + he_s), Du(x + he_s))$$
$$+ a(x, u(x + he_s), Du(x + he_s)) - a(x, u(x), Du(x + he_s))$$
$$+ a(x, u(x), Du(x + he_s)) - a(x, u(x), Du(x))$$
$$=: \mathcal{A}(h) + \mathcal{B}(h) + \mathcal{C}(h) \qquad (5.24)$$

with the obvious abbreviations (where, for ease of notation, we omit the x-argument). For later convenience, let us observe that $\mathcal{C}(h)$ may be rewritten as

$$\mathcal{C}(h) = \int_0^1 D_z a\big(x, u(x), Du(x) + t\tau_{s,h} Du(x)\big)\big) dt \, \tau_{s,h} Du(x)$$

$$=: \tilde{\mathcal{C}}(h)\tau_{s,h} Du(x) \, . \qquad (5.25)$$

Hence, equation (5.23) can be written as

$$\int_\Omega \left[\mathcal{A}(h) + \mathcal{B}(h) + \tilde{\mathcal{C}}(h)\tau_{s,h} Du\right] \cdot \tau_{s,h} Du \eta^2 \, dx$$

$$= -2 \int_\Omega \left[\mathcal{A}(h) + \mathcal{B}(h) + \tilde{\mathcal{C}}(h)\tau_{s,h} Du\right] \cdot \tau_{s,h} u \otimes D\eta \eta \, dx \, . \qquad (5.26)$$

In the next step we estimate the various terms arising in this identity. Note that the terms involving $\mathcal{B}(h)$ do not show up in the setting of statement (i).

Estimates for the integrals involving $\mathcal{A}(h)$. Using the Hölder continuity assumption (H3), Young's inequality, and the fact that $|h| < R/4$, we obtain for every $\varepsilon \in (0,1)$:

$$\int_\Omega |\mathcal{A}(h) \cdot \tau_{s,h} Du| \eta^2 \, dx$$

$$\leq 2L\omega_\alpha(|h|) \int_\Omega \left(1 + |Du(x+he)|\right) |\tau_{s,h} Du| \eta^2 \, dx$$

$$\leq \varepsilon \int_\Omega |\tau_{s,h} Du|^2 \eta^2 \, dx + 2L^2 \varepsilon^{-1} |h|^{2\alpha} \int_{B_{3R/4}(x_0)} \left(1 + |Du(x+he)|^2\right) dx$$

$$\leq \varepsilon \int_\Omega |\tau_{s,h} Du|^2 \eta^2 \, dx + 2L^2 \varepsilon^{-1} |h|^{2\alpha} \int_{B_R(x_0)} \left(1 + |Du|^2\right) dx.$$

Similarly, we infer for the second term

$$2 \int_\Omega |\mathcal{A}(h) \cdot \tau_{s,h} u \otimes D\eta| \eta \, dx$$

$$\leq c(R) L |h|^\alpha \int_\Omega \left(1 + |Du(x+he)|\right) |\tau_{s,h} u| \eta \, dx$$

$$\leq c(R) L |h|^{2\alpha} \int_{B_R(x_0)} \left(1 + |Du|^2\right) dx + c(R) L \int_{B_{3R/4}(x_0)} |\tau_{s,h} u|^2 \, dx.$$

Estimates for the integrals involving $\mathcal{B}(h)$. Applying (H3) and Young's inequality we find

$$\int_\Omega |\mathcal{B}(h) \cdot \tau_{s,h} Du| \eta^2 \, dx$$

$$\leq 2L \int_\Omega \left(1 + |Du(x+he)|\right) \omega_\alpha(|\tau_{s,h} u|) |\tau_{s,h} Du| \eta^2 \, dx$$

$$\leq \varepsilon \int_\Omega |\tau_{s,h} Du|^2 \eta^2 \, dx$$

$$+ 2L^2 \varepsilon^{-1} \int_{B_{3R/4}(x_0)} \left(1 + |Du(x+he)|^2\right) \omega_\alpha^2(|\tau_{s,h} u|) \, dx,$$

and, similarly, we obtain for the other term

$$2 \int_{\Omega} |\mathcal{B}(h) \cdot \tau_{s,h} u \otimes D\eta| \eta \, dx$$

$$\leq 4L \int_{\Omega} (1 + |Du(x+he)|) \omega_\alpha(|\tau_{s,h}u|) |\tau_{s,h}u| |D\eta| \eta \, dx$$

$$\leq c(R)L \int_{B_{3R/4}(x_0)} (1 + |Du(x+he)|^2) \omega_\alpha^2(|\tau_{s,h}u|) \, dx$$

$$+ c(R)L \int_{B_{3R/4}(x_0)} |\tau_{s,h}u|^2 \, dx .$$

Estimates for the integrals involving $\tilde{\mathcal{C}}(h)$. Keeping in mind the boundedness and ellipticity conditions (H1) and (H2) on $D_z a(\cdot, \cdot, \cdot)$, we easily check that $\tilde{\mathcal{C}}(h)$ from (5.25) is elliptic and bounded from above by L, see also Remark 5.1 (iii). Using the upper bound, we compute in exactly the same way as in the estimates above:

$$2 \int_{\Omega} |\tilde{\mathcal{C}}(h)\tau_{s,h} Du \cdot \tau_{s,h} u \otimes D\eta| \eta \, dx$$

$$\leq \varepsilon \int_{\Omega} |\tau_{s,h} Du|^2 \eta^2 \, dx + c(R)L^2 \varepsilon^{-1} \int_{B_{3R/4}(x_0)} |\tau_{s,h}u|^2 \, dx .$$

Finally, the ellipticity of $\tilde{\mathcal{C}}(h)$ is used to estimate

$$\int_{\Omega} \tilde{\mathcal{C}}(h)\tau_{s,h} Du \cdot \tau_{s,h} Du \, dx \geq \int_{\Omega} |\tau_{s,h} Du|^2 \eta^2 \, dx .$$

Collecting all estimates for the terms in (5.26), choosing $\varepsilon = 1/4$ and taking advantage of the estimate

$$\int_{B_{3R/4}(x_0)} |\tau_{s,h}u|^2 \, dx \leq c(R)|h|^{2\alpha} \int_{B_R(x_0)} |Du|^2 \, dx$$

(which follows from $|h| < R/4$ and Lemma 1.46), we obtain both assertions of the proposition. □

Remark 5.16 If the vector field a does not depend on the u-variable and is even Lipschitz continuous with respect to the x-variable (that is, $\alpha = 1$), then the previous proposition combined with Lemma 1.48 on difference quotients implies local $W^{2,2}$-regularity of weak solutions, a fact which was known for a long time.

In order to explain the strategy of the dimension reduction (and to simplify matters) we start by considering a vector field of the form $a(x, z)$ (as in the statement (i) above). Hence, as already observed in the previous remark, Lipschitz continuity of a with respect to the x-variable implies local $W^{2,2}$-regularity of weak solutions. Therefore, via the classical measure density result from Proposition 1.76 and the characterization of the singular set in Theorem 5.9, we find

$$\dim_{\mathcal{H}}(\mathrm{Sing}_0(Du)) \leq n - 2 \,.$$

If instead only Hölder continuity on the vector field a with respect to the x-variable is available, then the basic idea is that still a fractional differentiability of the gradient may be established, which in turn provides an estimate on the Hausdorff dimension of the singular set. This suggests that the regularity of the vector field a is not only related to the regularity of the solutions (in the sense that Du is locally Hölder continuous with the same exponent as the one of a), but also to the (maximal) Hausdorff dimension of the singular set, both according to the basic intuition which tells that an irregularity of the vector field a is an obstruction to regularity of weak solutions. This strategy for the dimension reduction was first accomplished by Mingione [62, 63] and is now given in a detailed way, first for vector fields a that are independent of the u-variable and then for the general case.

Systems without u-dependence We first state a consequence of Proposition 5.15 which, for arbitrary exponents $\alpha \in (0,1)$, still guarantees differentiability of Du in a fractional sense, even though the existence of second order derivatives of u cannot be ensured.

Proposition 5.17 *Let $u \in W^{1,2}(\Omega, \mathbb{R}^N)$ be a weak solution to the system (5.1) with a vector field $a \colon \Omega \times \mathbb{R}^{Nn} \to \mathbb{R}^{Nn}$ (not depending on the u-variable) which satisfies the assumptions (H1), (H2), and (H3). Then we have $Du \in W^{\beta,2}_{\mathrm{loc}}(\Omega, \mathbb{R}^{Nn})$ for all $\beta < \alpha$.*

Proof This is a consequence of Proposition 5.15, combined with a standard covering argument and Lemma 1.50. ☐

This proposition already leads to the desired dimension reduction for the set of (interior) singular points of Du, namely that its Hausdorff dimension does not exceed $n - 2\alpha$. However, for the sake of completeness, we also want to discuss a slight improvement of the previous fractional differentiability result, which is based on the use of Gehring's lemma.

Proposition 5.18 *Let $u \in W^{1,2}(\Omega, \mathbb{R}^N)$ be a weak solution to the system (5.1) with a vector field $a \colon \Omega \times \mathbb{R}^{Nn} \to \mathbb{R}^{Nn}$ (not depending on the u-variable) which satisfies the assumptions* (H1), (H2), *and* (H3). *Then there exists $\bar{p} = \bar{p}(n, N, L) > 2$ such that we have $Du \in W_{\text{loc}}^{\beta, \bar{p}}(\Omega, \mathbb{R}^{Nn})$ for all $\beta < \alpha$.*

Proof We proceed as in [63] and want to show that h-independent reverse Hölder inequalities for the gradients of the functions

$$w_h := |h|^{-\alpha} \tau_{s,h} u$$

are available, which in turn, via Gehring's lemma applied to Dw_h, provide the desired result. The strategy of proof is to adapt and refine the estimates from the proof of the previous Proposition 5.15, from where we also take the notation.

We start by fixing an arbitrary ball $B_{3R}(x_0) \subset \Omega$ and by taking a standard cut-off function $\eta \in C_0^\infty(B_{3R/4}(x_0), [0,1])$ which satisfies $\eta \equiv 1$ in $B_{R/2}(x_0)$ and $|D\eta| \leq 8/R$. We further consider $h \in \mathbb{R} \setminus \{0\}$ with $|h| < R/4$ and $s \in \{1, \ldots, n\}$. Since the vector field u does not depend explicitly on the u-variable, the terms involving $\mathcal{B}(h)$ all vanish, and we find analogously to the derivation of (5.26)

$$\int_\Omega \left[\mathcal{A}(h) + \tilde{\mathcal{C}}(h) \tau_{s,h} Du \right] \cdot \tau_{s,h} Du\, \eta^2 \, dx$$

$$= -2 \int_\Omega \left[\mathcal{A}(h) + \tilde{\mathcal{C}}(h) \tau_{s,h} Du \right] \cdot \left(\tau_{s,h} u - (\tau_{s,h} u)_{B_R(x_0)} \right) \otimes D\eta\, \eta \, dx \,,$$

by using the modified test function $\varphi := \tau_{s,-h}(\eta^2(\tau_{s,h} u - (\tau_{s,h} u)_{B_R(x_0)}))$ for the weak formulation of system (5.1). After dividing by $|h|^{2\alpha}$, we can express this identity in terms of w_h as

$$\int_\Omega \left[|h|^{-\alpha} \mathcal{A}(h) + \tilde{\mathcal{C}}(h) Dw_h \right] \cdot Dw_h\, \eta^2 \, dx$$

$$= -2 \int_\Omega \left[|h|^{-\alpha} \mathcal{A}(h) + \tilde{\mathcal{C}}(h) Dw_h \right] \cdot \left(w_h - (w_h)_{B_R(x_0)} \right) \otimes D\eta\, \eta \, dx \,. \tag{5.27}$$

Since $\tilde{\mathcal{C}}(h)$ is elliptic and bounded from above by L, cf. Remark 5.1 (iii), we can apply first Young's inequality and then the Sobolev–Poincaré inequality in a standard way to get

$$\fint_{B_{R/2}(x_0)} |Dw_h|^2\, dx$$

$$\leq c(L)R^{-2} \fint_{B_R(x_0)} |w_h - (w_h)_{B_R(x_0)}|^2\, dx + c \fint_{B_R(x_0)} |h|^{-2\alpha} |\mathcal{A}(h)|^2\, dx$$

$$\leq c(n, N, L) \left(\fint_{B_R(x_0)} |Dw_h|^{\frac{2n}{n+2}}\, dx \right)^{\frac{n+2}{n}} + c \fint_{B_R(x_0)} |h|^{-2\alpha} |\mathcal{A}(h)|^2\, dx$$

$$(5.28)$$

We next observe that, in view of the higher integrability result for Du from Lemma 5.3, the functions $\mathcal{A}(h)$ belong in fact to $L^p(B_R(x_0), \mathbb{R}^{Nn})$, for some $p > 2$ depending only on n, N, and L. Using also (H3) and the definition of $\mathcal{A}(h)$ in (5.24), we thus find the estimate

$$\left(\fint_{B_R(x_0)} |h|^{-p\alpha} |\mathcal{A}(h)|^p\, dx \right)^{\frac{2}{p}} \leq c(n, N, L) \fint_{B_{3R}(x_0)} (1 + |Du|)^2\, dx. \quad (5.29)$$

Taking into account that the ball $B_R(x_0)$ was arbitrary with $B_{3R}(x_0) \subset \Omega$ and that all constants in the previous inequalities are independent of h, we are in the position to apply Gehring's Theorem 1.22 with $\sigma = 1/2$ and $m = n/(n+2)$. Consequently, we find $\bar{p} \in (2, p)$ depending only on n, N, and L (and in particular independent of h) such that we have, again for any ball $B_{3R}(x_0) \subset \Omega$, the estimate

$$\left(\fint_{B_{R/2}(x_0)} |Dw_h|^{\bar{p}}\, dx \right)^{\frac{2}{\bar{p}}} \leq c(n, N, L) \fint_{B_R(x_0)} |Dw_h|^2\, dx$$

$$+ c(n, N, L) \left(\fint_{B_R(x_0)} |h|^{-\bar{p}\alpha} |\mathcal{A}(h)|^{\bar{p}}\, dx \right)^{\frac{2}{\bar{p}}}.$$

Finally, relying on the definition of w_h, the estimates (5.29) and the inequality from Proposition 5.15 (i), we arrive at

$$\int_{B_{R/2}(x_0)} |\tau_{s,h} Du|^{\bar{p}}\, dx \leq c(n, N, L, R) \left(\int_{B_{3R}(x_0)} (1 + |Du|)^2\, dx \right)^{\frac{\bar{p}}{2}} |h|^{\bar{p}\alpha}$$

for any $s \in \{1, \dots, n\}$. At this point the conclusion follows as in the proof of previous proposition, via a covering argument and the application of Lemma 1.50. $\qquad\qquad\square$

Now we have all ingredients at our disposal in order to prove an upper
bound for the Hausdorff dimension of the singular set of Du. This was
accomplished first by Mingione in [62, 63], and in the present situation, where
of vector field a does not depend on the solution itself explicitly, the proof
already exhibits all underlying ideas.

Theorem 5.19 (Mingione)   *Let $u \in W^{1,2}(\Omega, \mathbb{R}^N)$ be a weak solution to
the system* (5.1) *with a vector field $a\colon \Omega \times \mathbb{R}^{Nn} \to \mathbb{R}^{Nn}$ (not depending on
the u-variable) which satisfies the assumptions* (H1), (H2), (H3), *and* (H4).
*Then we have $\mathcal{H}^{n-2\alpha}(\mathrm{Sing}_0(Du)) = 0$ and moreover, there exists a positive
number $\sigma > 0$, depending only on n, N, L, and α such that*

$$\dim_{\mathcal{H}}(\mathrm{Sing}_0(Du)) \leq n - 2\alpha - \sigma. \tag{5.30}$$

Proof We proceed in two steps and first argue that the assertion (5.30) is true
for $\sigma = 0$. According to Proposition 5.17, there holds $Du \in W^{\beta,2}_{\mathrm{loc}}(\Omega, \mathbb{R}^{Nn})$
for all $\beta < \alpha$, and since all arguments are local, we may actually suppose
$Du \in W^{\beta,2}(\Omega, \mathbb{R}^{Nn})$ for all $\beta < \alpha$ (otherwise, we need to work on open
subsets $\Omega_k \Subset \Omega$ with $\cup_{k \in \mathbb{N}} \Omega_k = \Omega$). Since the set of interior singular points
of Du was characterized in Theorem 5.9 by $\mathrm{Sing}_0(Du) = \Sigma_1 \cup \Sigma_2$ with

$$\Sigma_1 = \left\{ x_0 \in \Omega \colon \liminf_{\varrho \searrow 0} \fint_{\Omega(x_0,\varrho)} |Du - (Du)_{\Omega(x_0,\varrho)}|^2 \, dx > 0 \right\},$$

$$\Sigma_2 = \left\{ x_0 \in \Omega \colon \limsup_{\varrho \searrow 0} |(Du)_{\Omega(x_0,\varrho)}| = \infty \right\},$$

the measure density result from Proposition 1.76 allows to conclude

$$\mathcal{H}^{n-2\beta+\delta}(\Sigma_1) = 0 \quad \text{and} \quad \mathcal{H}^{n-2\beta+\delta}(\Sigma_2) = 0,$$

for all $\beta < \alpha$ and $\delta > 0$. By definition of the Hausdorff dimension, this yields

$$\dim_{\mathcal{H}}(\mathrm{Sing}_0(Du)) \leq n - 2\alpha.$$

In order to prove in a second step the full statement of the theorem, we employ
the refined fractional differentiability statement from Proposition 5.18 instead
of Proposition 5.17, and for simplicity we assume global higher integrability
$Du \in W^{\beta,\bar{p}}(\Omega, \mathbb{R}^{Nn})$ for all $\beta < \alpha$ and some $\bar{p}$ depending only on n, N,
and L. Via Jensen's inequality, we have the inclusion

$$\Sigma_1 \subset \left\{ x_0 \in \Omega \colon \liminf_{\varrho \searrow 0} \fint_{\Omega(x_0,\varrho)} |Du - (Du)_{\Omega(x_0,\varrho)}|^{\bar{p}} \, dx > 0 \right\},$$

and analogously to above we obtain $\mathcal{H}^{n-\bar{p}\beta+\delta}(\mathrm{Sing}_0(Du)) = 0$ via Proposition 1.76, for all $\beta < \alpha$ and $\delta > 0$. With $\bar{p} > 2$ we thus find a positive number σ, depending only on n, N, L, and α, such that the assertion (5.30) holds, and the proof of the theorem is complete. $\qquad\square$

Systems with u-dependence We finally pass to general coefficients $a(x, u, z)$ possibly depending also on the u-variable. In order to follow the line of arguments from above, we have to investigate the regularity of the map $x \mapsto (x, u(x))$. For this reason it seems unlikely to obtain a dimension reduction result as above without requiring some better regularity properties of the weak solution u. Let us assume that u is a priori λ-Hölder continuous everywhere. In this situation, $x \mapsto (x, u(x))$ is also Hölder continuous and the previous arguments still apply. However, some modifications are necessary due to the fact that $x \mapsto (x, u(x))$ is not Hölder continuous with exponent α, but only with exponent $\alpha\lambda$, because of the presence of u. One possibility to gain nevertheless the fractional Sobolev space regularity as in Proposition 5.17 relies on a subtle iteration argument. The key element here is an interpolation result due to Campanato [10] (and related to the classical Gagliardo–Nirenberg inequality).

Theorem 5.20 (Campanato) *Let $B_R(x_0) \subset \mathbb{R}^n$, $\lambda, \theta \in (0, 1]$ and $p \in (1, \infty)$ such that $p\theta < n$. If $u \in C^{0,\lambda}(B_R(x_0), \mathbb{R}^N) \cap W^{1+\theta,p}(B_R(x_0), \mathbb{R}^{nN})$, then we have*

$$Du \in L^q(B_{R/2}(x_0), \mathbb{R}^{nN}) \qquad \text{for all } q < \frac{np(1 + \theta)}{n - p\theta\lambda},$$

with

$$\int_{B_{R/2}(x_0)} |Du|^q \, dx \le c$$

for a constant c depending only on n, N, p, θ, λ, q, R, $\|u\|_{W^{1+\theta,p}(B_R(x_0), \mathbb{R}^N)}$, and $[u]_{C^{0,\lambda}(B_R(x_0), \mathbb{R}^N)}$.

Proposition 5.21 *Let $u \in C^{0,\lambda}(\Omega, \mathbb{R}^N) \cap W^{1,2}(\Omega, \mathbb{R}^N)$ be a weak solution to the system (5.1) with a vector field $a \colon \Omega \times \mathbb{R}^N \times \mathbb{R}^{Nn} \to \mathbb{R}^{Nn}$ which satisfies the assumptions (H1), (H2), and (H3). Then we have $Du \in W^{\beta,2}_{\mathrm{loc}}(\Omega, \mathbb{R}^{Nn})$ for all $\beta < \alpha$.*

Proof The proof of the proposition is carried out by an iteration. One starts with the fractional Sobolev estimate provided by Proposition 5.17, then gains some higher integrability of Du which in turn allows us to reenter in the first fractional Sobolev estimate and to improve the fractional differentiability. In this way, step by step, the regularity of Du is improved up to local $W^{\beta,2}$ regularity for any given $\beta < \alpha$. More precisely, one proceeds as follows.

We take a ball $B_{\varrho_k}(x_0) \Subset \Omega$ for some radius $\varrho_k < 1$ and we suppose that for some $\alpha_k < (0, \alpha)$ we have the following condition:

(i)$_k$ For $s \in \{1, \ldots, n\}$ and $h \in \mathbb{R}$ with $|h| < \varrho_k/4$ we have an estimate for finite differences of Du of the form

$$\int_{B_{\varrho_k}(x_0)} |\tau_{s,h} Du|^2 \, dx \leq c_k |h|^{2\alpha_k}$$

for some constant c_k (which is independent of h).

Then we can deduce the following chain of implications:

(ii)$_k$ As a consequence of Lemma 1.50, the assumption (i)$_k$ implies $Du \in W^{\gamma\alpha_k,2}(B_{\varrho_k/2}(x_0), \mathbb{R}^{Nn})$ for every $\gamma \in (0,1)$, with an estimate of the fractional Sobolev semi-norm of Du via

$$\int_{B_{\varrho_k/2}(x_0)} \int_{B_{\varrho_k/2}(x_0)} \frac{|Du(x) - Du(y)|^2}{|x-y|^{n+2\gamma\alpha_k}} \, dx \, dy \leq c,$$

with c depending only on n, N, γ, α_k, ϱ_k, c_k, and $\|Du\|_{L^2(\Omega,\mathbb{R}^{Nn})}$.

(iii)$_k$ Campanato's interpolation Theorem 5.20 then provides the higher integrability result $Du \in L^{2+2\alpha_k}(B_{\varrho_k/4}(x_0), \mathbb{R}^{Nn})$, since

$$2 + 2\alpha_k < \frac{2n(1+\gamma\alpha_k)}{n - 2\gamma\alpha_k\lambda},$$

whenever the number γ in (ii)$_k$ is chosen sufficiently close to 1 (for instance, $\gamma > (1+2\lambda/n)^{-1}$ is sufficient). Moreover, we have the estimate

$$\int_{B_{\varrho_k/4}(x_0)} |Du|^{2+2\alpha_k} \, dx \leq c$$

with a constant c depending only on n, N, λ, α_k, ϱ_k, c_k, $\|Du\|_{L^2(\Omega,\mathbb{R}^{Nn})}$, and $[u]_{C^{0,\lambda}(\Omega,\mathbb{R}^N)}$.

(iv)$_k$ Next the second term on the right-hand side of the inequality in Proposition 5.15 (ii) is estimated via the fact that $\omega_\alpha(t) \leq t^\alpha$ holds for all $t \geq 0$, Hölder's inequality and by taking into account both the higher integrability for Du from (iii)$_k$ and the Hölder continuity of u:

$$\int_{B_{3\varrho_k/16}(x_0)} |Du(x+he_s)|^2 |\tau_{s,h} u(x)|^{2\alpha} \, dx$$

$$\leq \left(\int_{B_{\varrho_k/4}(x_0)} |Du|^{2+2\alpha_k} \, dx \right)^{\frac{1}{1+\alpha_k}}$$

$$\times \left(\int_{B_{3\varrho_k/16}(x_0)} |\tau_{s,h}u(x)|^{2\alpha\frac{1+\alpha_k}{\alpha_k}} \, dx \right)^{\frac{\alpha_k}{1+\alpha_k}}$$

$$= \left(\int_{B_{\varrho_k/4}(x_0)} |Du|^{2+2\alpha_k} \, dx \right)^{\frac{1}{1+\alpha_k}}$$

$$\times \left(\int_{B_{3\varrho_k/16}(x_0)} |\tau_{s,h}u(x)|^{2(1+\alpha_k)+2(\alpha-\alpha_k)\frac{1+\alpha_k}{\alpha_k}} \, dx \right)^{\frac{\alpha_k}{1+\alpha_k}}$$

$$\leq c([u]_{C^{0,\lambda}(\Omega,\mathbb{R}^N)}) |h|^{2\alpha_k+2(\alpha-\alpha_k)\lambda} \int_{B_{\varrho_k/4}(x_0)} |Du|^{2+2\alpha_k} \, dx$$

(with h, s as in Proposition 5.15 and with R replaced by $\varrho_k/4$). In order to obtain the last inequality we have on the one hand employed Lemma 1.46 to estimate the integral over $|\tau_{s,h}u(x)|^{2+2\alpha_k}$ by the corresponding integral over $|Du|^{2+2\alpha_k}$ (with prefactor $|h|^{2+2\alpha_k}$), and on the other hand we have involved the λ-Hölder continuity of u to estimate the remaining differences $|\tau_{s,h}u(x)|^{2(1+\alpha_k)(\alpha-\alpha_k)/\alpha_k}$ in terms of $|h|^{\lambda 2(1+\alpha_k)(\alpha-\alpha_k)/\alpha_k}$.

(v)$_k$ Plugging the estimates in (iv)$_k$ and (iii)$_k$ into the inequality of Proposition 5.15 (ii), we find that, for every $s \in \{1,\ldots,n\}$ and $h \in \mathbb{R}$ with $|h| < \varrho_k/16$, the following estimate holds for finite differences of Du:

$$\int_{B_{\varrho_k/8}(x_0)} |\tau_{s,h}Du|^2 \, dx \leq c|h|^{2\alpha_k+2(\alpha-\alpha_k)\lambda},$$

and the constant c depends only on n, N, λ, α_k, ϱ_k, c_k, $\|Du\|_{L^2(\Omega,\mathbb{R}^{Nn})}$, and $\|u\|_{C^{0,\lambda}(\Omega,\mathbb{R}^{Nn})}$.

Now we observe that, for suitable choices of ϱ_{k+1} and α_{k+1}, the estimate (v)$_k$ is of the form (i)$_{k+1}$. Therefore, the proof of the proposition is concluded as follows: we fix $\beta < \alpha$ and an initial ball $B_\varrho(x_0) \Subset \Omega$. Then we define the sequence of radii $(\varrho_k)_{k\in\mathbb{N}}$ and of numbers $(\alpha_k)_{k\in\mathbb{N}}$ via

$$\varrho_k := 2^{-1}8^{-k+1}\varrho \quad \text{and} \quad \alpha_k := \alpha\lambda\sum_{\ell=0}^{k-1}(1-\lambda)^\ell$$

for $k \in \mathbb{N}$. Hence, the sequence $(\varrho_k)_{k\in\mathbb{N}}$ is decreasing in k, the sequence $(\alpha_k)_{k\in\mathbb{N}}$ is increasing in k with limit α, and moreover, α_{k+1} can recursively be defined via

$$\alpha_{k+1} = \alpha_k + (\alpha - \alpha_k)\lambda$$

for $k \in \mathbb{N}$, and this is exactly the formula appearing in the exponent of the estimate in $(v)_k$. Next, we note that after a finite number of steps we have $\alpha_{k_0} \in (\beta, \alpha)$ for some $k_0 \in \mathbb{N}$. Thus, following the iteration scheme described above we find $Du \in W^{\beta,2}(B_{\varrho_k}(x_0), \mathbb{R}^{Nn})$ (and from the statement $(v)_k$ we also see in terms of which quantities this norm is estimated). The final claim then follows from a standard covering argument.   □

Also in this setting, we still can slightly improve on the fractional differentiability result and gain a better integrability exponent.

Proposition 5.22 *Let* $u \in C^{0,\lambda}(\Omega, \mathbb{R}^N) \cap W^{1,2}(\Omega, \mathbb{R}^N)$ *be a weak solution to the system* (5.1) *with a vector field* $a: \Omega \times \mathbb{R}^N \times \mathbb{R}^{Nn} \to \mathbb{R}^{Nn}$ *which satisfies the assumptions* (H1), (H2), (H3). *Then there exists* $\bar{p} = \bar{p}(n, N, L, \lambda, \alpha) > 2$ *such that we have* $Du \in W^{\beta,\bar{p}}_{\mathrm{loc}}(\Omega, \mathbb{R}^{Nn})$ *for all* $\beta < \alpha$.

Proof The strategy of proof is exactly the same as for Proposition 5.18, from where we take the notation. Therefore, we only comment briefly on the necessary modifications, which are essentially caused be the presence of the integrals involving $\mathcal{B}(h)$, in order to justify the uniform higher integrability results for the functions $w_h := |h|^{-\alpha}\tau_{s,h}u$. The starting point for all computations is a variant of (5.27), which now, in presence of the u-dependency of the vector field a, is given by the identity

$$\int_\Omega \left[|h|^{-\alpha}\mathcal{A}(h) + |h|^{-\alpha}\mathcal{B}(h) + \tilde{\mathcal{C}}(h)Dw_h\right] \cdot Dw_h \eta^2 \, dx$$

$$= -2\int_\Omega \left[|h|^{-\alpha}\mathcal{A}(h) + |h|^{-\alpha}\mathcal{B}(h) + \tilde{\mathcal{C}}(h)Dw_h\right] \cdot \left(w_h - (w_h)_{B_R(x_0)}\right) \otimes D\eta\eta \, dx.$$

Proceeding for the terms involving $\mathcal{A}(h)$ and $\tilde{\mathcal{C}}(h)$ exactly as before, we here obtain

$$\fint_{B_{R/2}(x_0)} |Dw_h|^2 \, dx \leq c(n, N, L)\left(\fint_{B_R(x_0)} |Dw_h|^{\frac{2n}{n+2}} \, dx\right)^{\frac{n+2}{n}}$$

$$+ c\fint_{B_R(x_0)} |h|^{-2\alpha}|\mathcal{A}(h)|^2 \, dx + c\fint_{B_R(x_0)} |h|^{-2\alpha}|\mathcal{B}(h)|^2 \, dx,$$

which is completely analogous to the estimate (5.28). Therefore, it only remains to justify the higher integrability of the integrand involving $\mathcal{B}(h)$ on the right-hand side with a suitable estimate which is in particular independent of h. To this end, we may take advantage of the fact that $Du \in L^q_{\mathrm{loc}}(\Omega, \mathbb{R}^{Nn})$ for some exponent $q > 2 + 2\alpha$, which is a consequence of Theorem 5.20 combined with the outcome of Proposition 5.21 (for β sufficiently close to α).

With the definition of $\mathcal{B}(h)$ from (5.24) and the assumption (H3) on Hölder continuity of a with respect to the first and second variable, we initially find

$$\fint_{B_R(x_0)} |\mathcal{B}(h)|^2\, dx \leq c(L) \fint_{B_R(x_0)} \left(1 + |Du(x + he_s)|\right)^2 |\tau_{h,s} u|^{2\alpha}\, dx\,,$$

and then the application of Hölder's inequality and Lemma 1.46 shows the boundedness of the right-hand side:

$$\fint_{B_R(x_0)} \left(1 + |Du(x + he_s)|\right)^2 |\tau_{h,s} u|^{2\alpha}\, dx$$

$$\leq c(L)\left(\fint_{B_R(x_0)} \left(1+|Du(x + he_s)|\right)^{2+2\alpha}\, dx\right)^{\frac{1}{1+\alpha}} \left(\fint_{B_R(x_0)} |\tau_{h,s} u|^{2+2\alpha}\, dx\right)^{\frac{\alpha}{1+\alpha}}$$

$$\leq c(n, L)|h|^{2\alpha} \fint_{B_{2R}(x_0)} \left(1 + |Du|\right)^{2+2\alpha}\, dx\,. \tag{5.31}$$

Moreover, the same line of arguments leads to a higher integrability estimate

$$\fint_{B_R(x_0)} \left(|h|^{-\alpha}|\mathcal{B}(h)|\right)^{\frac{2q}{2+2\alpha}}\, dx \leq c(n, L) \fint_{B_{2R}(x_0)} \left(1 + |Du|\right)^q\, dx \tag{5.32}$$

which should be compared to (5.29). With these estimates at hand, we are again in the position to apply Gehring's Theorem 1.22 with $\sigma = 1/2$ and $m = n/(n+2)$, which yields a number $\bar{p} \in (2, \min\{p, 2q/(2+2\alpha)\})$ depending only on n, N, L, λ, and α (and in particular independent of h) such that we have, again for any ball $B_{3R}(x_0) \subset \Omega$, the estimate

$$\left(\fint_{B_{R/2}(x_0)} |Dw_h|^{\bar{p}}\, dx\right)^{\frac{2}{\bar{p}}} \leq c \fint_{B_R(x_0)} |Dw_h|^2\, dx$$

$$+ c\left(\fint_{B_R(x_0)} \left(|h|^{-\alpha}|\mathcal{A}(h)| + |h|^{-\alpha}|\mathcal{B}(h)|\right)^{\bar{p}}\, dx\right)^{\frac{2}{\bar{p}}},$$

with the constant c still depending only on n, N, and L. Finally, relying on the definition of w_h, the estimates (5.29) and (5.32), and the inequality from Proposition 5.15 (ii) combined with (5.31), we arrive at

$$\int_{B_{R/2}(x_0)} |\tau_{s,h} Du|^{\bar{p}}\, dx \leq c\left(\int_{B_{3R}(x_0)} \left(1 + |Du|\right)^{2+2\alpha}\, dx\right)^{\frac{\bar{p}}{2}} |h|^{\bar{p}\alpha}$$

for any $s \in \{1, \ldots, n\}$ and a constant c depending only on n, N, L, λ, α, R, and $[u]_{C^{0,\lambda}(\Omega, \mathbb{R}^N)}$. At this point the conclusion follows as before, via a covering argument and the application of Lemma 1.50.   □

Remark 5.23 If, in the setting of Proposition 5.22, the weak solution is not assumed to be Hölder continuous, then the interpolation Theorem 5.20 cannot be used. However, one can still prove the fractional differentiability result $Du \in W^{\beta, \bar{p}}_{\text{loc}}(\Omega, \mathbb{R}^{Nn})$ for some $\beta > 0$ (depending only on the data), cf. [62, Proposition 5.3].

With this fractional Sobolev regularity results at hand, the dimension reduction of the singular set of Du follows exactly as before from the characterization of $\text{Sing}_0(Du)$ stated in Theorem 5.9 and the measure density result in Proposition 1.76, cf. [62, Theorem 1.1].

Theorem 5.24 (Mingione) *Let $u \in C^{0,\lambda}(\Omega, \mathbb{R}^N) \cap W^{1,2}(\Omega, \mathbb{R}^N)$ be a weak solution to the system (5.1) with a vector field $a: \Omega \times \mathbb{R}^N \times \mathbb{R}^{Nn} \to \mathbb{R}^{Nn}$ which satisfies the assumptions (H1), (H2), (H3), and (H4). Then we have $\mathcal{H}^{n-2\alpha}(\text{Sing}_0(Du)) = 0$ and moreover, there exists a positive number $\sigma > 0$, depending only on n, N, L, λ, and α, such that*

$$\dim_{\mathcal{H}}(\text{Sing}_0(Du)) \leq n - 2\alpha - \sigma.$$

Remarks 5.25

(i) The proof is self-contained and involves only PDE arguments, but it does not rely on potential estimates.

(ii) This is the best estimate known for the Hausdorff dimension of the singular set in such a general quasilinear setting. Although we cannot rule out the possibility that the dependence on the Hölder exponent α is purely technical, it is believed that it is a structural feature.

(iii) The statement of the theorem remains true if the assumption $u \in C^{0,\lambda}(\Omega, \mathbb{R}^N)$ is replaced by the assumption $\dim_{\mathcal{H}}(\text{Sing}_\lambda(u)) \leq n - 2\alpha$ for some $\lambda > 0$. In this case we may restrict the analysis of Du to the regular set $\text{Reg}_\lambda(u)$. This is sufficient to infer the same conclusion since all estimates are local and since we are only interested in sets of Hausdorff dimension not less than $n - 2\alpha$. Such a weaker assumption (more precisely, that $\dim_{\mathcal{H}}(\text{Sing}_\lambda(u)) \leq n-2$ holds for $\lambda = 1-(n-2)/2$) is known to be true for weak solutions to systems of the form (5.1) under the above assumptions and the additional restriction to low dimensions $n \leq 4$, see Theorem 5.30.

We conclude the dimension reduction of the singular set of Du with a final result, which is related to the previous remark and deals with general quasilinear system, when no a priori Hölder continuity of the weak solution is assumed. As observed by Mingione [62], it turns out that the Hausdorff dimension of $\text{Sing}_0(Du)$ is still strictly bounded by the space dimension n, and hence, the statement $\mathcal{L}^n(\mathrm{Sing}_0(Du)) = 0$ obtained in Theorem 5.9 is never optimal.

Theorem 5.26 (Mingione) *Let $u \in W^{1,2}(\Omega, \mathbb{R}^N)$ be a weak solution to the system (5.1) with a vector-field $a\colon \Omega \times \mathbb{R}^N \times \mathbb{R}^{Nn} \to \mathbb{R}^{Nn}$ which satisfies the assumptions (H1), (H2), (H3), and (H4). Then we have the following estimate*

$$\dim_{\mathcal{H}}(\mathrm{Sing}_0(Du)) \le n - \min\{2\alpha, p - 2\},$$

where $p(n, N, L) > 2$ is the higher integrability exponent from Lemma 5.3.

Proof In view of the higher integrability result $u \in W^{1,p}_{\mathrm{loc}}(\Omega, \mathbb{R}^N)$ from Lemma 5.3, we can estimate both integrals appearing on the right-hand side of Proposition 5.15 (ii) in terms of $|h|^{\min\{2\alpha, p-2\}}$. This (uniform) estimate on finite differences of Du implies in turn $Du \in W^{\beta,2}_{\mathrm{loc}}(\Omega, \mathbb{R}^{Nn})$ for every $\beta < \min\{2\alpha, p - 2\}$ via Lemma 1.50. At this point, we may argue exactly as in the first part of the proof of Theorem 5.19 and arrive at the desired bound $n - \min\{2\alpha, p - 2\}$ for the Hausdorff dimension of $\mathrm{Sing}_0(Du)$. $\qquad\square$

5.3.2 *Bounds in Low Dimensions*

For general quasilinear systems (5.1), where no additional structure assumption on the vector-field a or no a priori regularity of the weak solution u is available, the best result for the Hausdorff dimension of $\mathrm{Sing}_0(Du)$, which has so far been achieved, is indeed the one in Theorem 5.26. Hence, the only bound for $\dim_{\mathcal{H}}(\mathrm{Sing}_0(Du))$ is just below the space dimension n. Better estimates on the Hausdorff dimension may be established, when the singular set $\mathrm{Sing}_0(u)$ of weak solutions $u \in W^{1,2}(\Omega, \mathbb{R}^N)$ to the quasilinear system (5.1) is investigated, under the restriction $n \le 4$ on the dimension. Since the singular set of the solution itself rather than its gradient is considered, we can work under slightly weaker assumptions than before, and in addition to (H1) and (H2), we now suppose

(H3') a is continuous with respect to (x, u) with

$$|a(x, u, z) - a(\tilde{x}, \tilde{u}, z)| \le 2L\omega\big(|x - \tilde{x}| + |u - \tilde{u}|\big)(1 + |z|),$$

for all $x, \tilde{x} \in \Omega$, $u, \tilde{u} \in \mathbb{R}^N$ and $z \in \mathbb{R}^{Nn}$. Here, $\omega\colon \mathbb{R}^+ \to [0, 1]$ is again a non-decreasing and concave modulus of continuity.

 We now show partial regularity of u outside of a set of Hausdorff dimension $n-2$. The strategy of proof complements the techniques presented in Sect. 4.3, even though it shares the crucial features of the direct approach from Sect. 4.3.3. The essential ingredient is again a comparison argument, but this time it does not involve the solution of a suitably linearized problem, but the solution of a nonlinear comparison system, where the (x, u)-variables in the vector field a are kept fixed and which for this reason is often referred to as the "frozen system".

Comparison estimates   We start by deriving the aforementioned estimates for the nonlinear comparison system. For this purpose, we study weak solutions $v \in W^{1,2}(B_R(x_0), \mathbb{R}^N)$ to a quasilinear elliptic system of the form

$$\operatorname{div} a_0(Dv) = 0 \quad \text{in } B_R(x_0), \tag{5.33}$$

where $B_R(x_0)$ is a ball in $\mathbb{R}^n$ and with a vector field $a_0 \colon \mathbb{R}^{Nn} \to \mathbb{R}^{Nn}$ which depends exclusively on the gradient variable and which satisfies the growth and ellipticity assumptions (H1) and (H2).

Remark 5.27 For such nonlinear systems the Browder–Minty Theorem A.12 ensures the existence of a (unique) weak solution in the Dirichlet class $u_0 + W_0^{1,2}(B_R(x_0), \mathbb{R}^N)$, for arbitrary prescribed boundary values $u_0 \in W^{1,2}(B_R(x_0), \mathbb{R}^N)$ (and actually also for every inhomogeneity in the space $(W_0^{1,2}(B_R(x_0), \mathbb{R}^N))^*$). In fact, the hypotheses of Theorem A.12 are satisfied in the reflexive, separable Banach space $W_0^{1,2}(B_R(x_0), \mathbb{R}^N)$, with the operator $A \colon W_0^{1,2}(B_R(x_0), \mathbb{R}^N) \to (W_0^{1,2}(B_R(x_0), \mathbb{R}^N))^*$ given by

$$\langle w, A(v) \rangle_{W_0^{1,2}(B_R(x_0),\mathbb{R}^N),(W_0^{1,2}(B_R(x_0),\mathbb{R}^N))^*} := \int_{B_R(x_0)} a_0(Du_0 + Dv) \cdot Dw \, dx$$

for all $v, w \in W_0^{1,2}(\Omega, \mathbb{R}^N)$ and with $F = 0$. In view of (H1), A is bounded and continuous, and in view of (H2), it is also strictly monotonic and coercive.

Similarly as in the case of linear systems in Sect. 4.2.2, we now prove decay estimates. However, due to the nonlinearity of the vector field a_0 the proof is slightly more technical than in the linear setting and requires in particular the use of Gehring's lemma (or alternatively a version of Widman's hole filling technique) to deduce a suitable higher integrability result. As a further consequence of the nonlinearity of a_0, the scaling of the Dirichlet energy on balls in terms of their radius differs from the scaling for linear systems, and this is the fundamental reason for the assumption $n \leq 4$ of low dimensions (which will become clear further below).

Lemma 5.28 (Decay estimates IV; Campanato)   *Let $v \in W^{1,2}(\Omega, \mathbb{R}^N)$ be a weak solution to the system (5.33) with a vector field $a_0 \colon \mathbb{R}^{Nn} \to \mathbb{R}^{Nn}$ satisfying (H1) and (H2). Then there exists a number $\varepsilon > 0$ depending only on n and L such that for all $B_r(x_0) \subset B_R(x_0) \subset \Omega$ we have*

$$\int_{B_r(x_0)} |Dv|^2 \, dx \leq c \left(\frac{r}{R} \right)^{\min\{n, 2+\varepsilon\}} \int_{B_R(x_0)} |Dv|^2 \, dx \tag{5.34}$$

and

$$\int_{B_r(x_0)} |Dv - (Dv)_{B_r(x_0)}|^2 \, dx \leq c\Big(\frac{r}{R}\Big)^{2+\varepsilon} \int_{B_R(x_0)} |Dv - (Dv)_{B_R(x_0)}|^2 \, dx,$$

$$(5.35)$$

with constants c depending only on n, N, and L.

Proof We here follow the exposition in [12]. In a first step, relying on the assumptions (H1) and (H2), we observe that all second order derivatives of v exist in $L^2_{\mathrm{loc}}(\Omega, \mathbb{R}^N)$, as a consequence of Proposition 5.15 and Lemma 1.48 (see also Remark 5.16). Moreover, the following Caccioppoli inequality

$$\int_{B_{\varrho/2}(y)} |D^2 v|^2 \, dx \leq c(L)\varrho^{-2} \int_{B_\varrho(y)} |Dv - (Dv)_{B_\varrho(y)}|^2 \, dx \qquad (5.36)$$

holds for any ball $B_\varrho(y) \subset \Omega$. Applying the Sobolev–Poincaré inequality (cf. Remarks 1.57) we then infer

$$\fint_{B_{\varrho/2}(y)} |D^2 v|^2 \, dx \leq c(n, N, L)\Big(\fint_{B_\varrho(y)} |D^2 v|^{\frac{2n}{n+2}} \, dx\Big)^{\frac{n+2}{n}},$$

and Gehring's Lemma (see Theorem 1.22) yields the existence of a number $p = p(L, n) > 2$ such that such v belongs to $W^{2,p}_{\mathrm{loc}}(\Omega, \mathbb{R}^N)$ and such that, in particular, the estimate

$$\Big(\fint_{B_{R/4}(x_0)} |D^2 v|^p \, dx\Big)^{\frac{2}{p}} \leq c(n, N, L) \fint_{B_{R/2}(x_0)} |D^2 v|^2 \, dx$$

is true. Via Jensen's inequality, this provides in the next step a decay estimate for second order derivatives of v. Indeed, for every $r \in (0, R/4)$, we find

$$\fint_{B_r(x_0)} |D^2 v|^2 \, dx \leq c(n)r^n \Big(\fint_{B_r(x_0)} |D^2 v|^p \, dx\Big)^{\frac{2}{p}}$$

$$\leq c(n)r^n \Big(\frac{r}{R}\Big)^{-\frac{2n}{p}} \Big(\fint_{B_{R/4}(x_0)} |D^2 v|^p \, dx\Big)^{\frac{2}{p}}$$

$$\leq c(n, N, L)\Big(\frac{r}{R}\Big)^\varepsilon \fint_{B_{R/2}(x_0)} |D^2 v|^2 \, dx,$$

where we have defined $\varepsilon := n(p - 2)/p > 0$. In order to derive the desired Campanato-type decay estimate (5.35) for Dv, we now proceed with essentially the same arguments as in the proof of Lemma 4.11. Using first the Poincaré inequality from Lemma 1.56, then the decay estimate for second order derivatives and finally the Caccioppoli inequality (5.36), we find

$$\int_{B_r(x_0)} |Dv - (Dv)_{B_r(x_0)}|^2 \, dx$$

$$\leq c(n, N) r^2 \int_{B_r(x_0)} |D^2 v|^2 \, dx$$

$$\leq c(n, N, L) r^2 \left(\frac{r}{R}\right)^\varepsilon \int_{B_{R/2}(x_0)} |D^2 v|^2 \, dx$$

$$\leq c(n, N, L) \left(\frac{r}{R}\right)^{2+\varepsilon} \int_{B_R(x_0)} |Dv - (Dv)_{B_r(x_0)}|^2 \, dx \, .$$

This is the desired estimate (5.35) for $r \in (0, R/4)$, while the analogous inequality for $r \in [R/4, R]$ is trivially satisfied with constant $4^{2+\varepsilon}$.

Concerning the Morrey-type decay estimate (5.34), we now distinguish only the cases $\varepsilon \in (0, n-2)$ and $\varepsilon \in (n-2, n)$ (by choosing ε possibly smaller). In the first case, in view of Jensen's inequality, there holds for all r, ϱ with $0 < r \leq \varrho \leq R$:

$$\int_{B_r(x_0)} |Dv|^2 \, dx$$

$$\leq c(n) r^n |(Dv)_{B_\varrho(x_0)}|^2 + 2 \int_{B_r(x_0)} |Dv - (Dv)_{B_\varrho(x_0)}|^2 \, dx$$

$$\leq c(n) \left(\frac{r}{\varrho}\right)^n \int_{B_\varrho(x_0)} |Dv|^2 \, dx + 2 \int_{B_\varrho(x_0)} |Dv - (Dv)_{B_\varrho(x_0)}|^2 \, dx$$

$$\leq c(n) \left(\frac{r}{\varrho}\right)^n \int_{B_\varrho(x_0)} |Dv|^2 \, dx + c(n, N, L) \left(\frac{\varrho}{R}\right)^{2+\varepsilon} \int_{B_R(x_0)} |Dv|^2 \, dx \, ,$$

where we have used the Campanato-type decay estimate (5.35), with r replaced by ϱ. The iteration Lemma B.3, applied with $\alpha_1 := n > 2 + \varepsilon =: \alpha_2$, $\kappa = 0$,

$$\phi(\varrho) := \int_{B_\varrho(x_0)} |Dv|^2 \, dx \quad \text{and} \quad A := R^{-2-\varepsilon} \int_{B_R(x_0)} |Dv|^2 \, dx \, ,$$

then yields

$$\int_{B_r(x_0)} |Dv|^2 \, dx \leq c(n, N, L) \left(\frac{r}{R}\right)^{2+\varepsilon} \int_{B_R(x_0)} |Dv|^2 \, dx \, ,$$

which is the desired estimate (5.34) in the case $\varepsilon \in (0, n-2)$. If, on the contrary, the case $\varepsilon \in (n-2, n)$ is considered, we observe from (5.35) that Dv belongs to the Campanato space $\mathcal{L}^{2,2+\varepsilon}_{\text{loc}}(\Omega, \mathbb{R}^{Nn})$. Thus, via the isomorphy of Campanato spaces and Hölder spaces given in Theorem 1.27, we conclude $Dv \in C^{0,\alpha}(\Omega, \mathbb{R}^{Nn})$ with Hölder exponent $\alpha = 1 - (n-\varepsilon)/2 > 0$. Furthermore, this isomorphy implies for all interior balls $B_R(x_0)$ the estimate

$$\sup_{B_{R/2}(x_0)} |Dv|^2 \le c(n,N) R^{-n} \left(\int_{B_{R/2}(x_0)} |Dv|^2 \, dx \right.$$

$$\left. + \sup_{\substack{y \in B_{R/2}(x_0), \\ \varrho > 0}} \left(\frac{\varrho}{R}\right)^{-2-\varepsilon} \int_{B_\varrho(y) \cap B_{R/2}(x_0)} \left| Dv - (Dv)_{B_\varrho(y) \cap B_{R/2}(x_0)} \right|^2 \, dx \right)$$

(for the precise dependence on R see Remark 1.28 (iv)). At this stage we observe that we can restrict ourselves to radii $\varrho < R - |y - x_0|$ in the supremum on the right-hand side (which in turn implies $B_\varrho(y) \subset B_R(x_0)$) since the estimate

$$\varrho^{-2-\varepsilon} \int_{B_\varrho(y) \cap B_{R/2}(x_0)} \left| Dv - (Dv)_{B_\varrho(y) \cap B_{R/2}(x_0)} \right|^2 \, dx$$

$$\le \left(\frac{R}{2}\right)^{-2-\varepsilon} \int_{B_{R/2}(x_0)} \left| Dv - (Dv)_{B_{R/2}(x_0)} \right|^2 \, dx$$

is satisfied for every radius $\varrho \ge R - |y - x_0| \ge R/2$. This allows us to continue estimating the supremum of $|Dv|$, and with the help of (5.35) we finally arrive at

$$\sup_{B_{R/2}(x_0)} |Dv|^2 \le c(n,N,L) R^{-n} \int_{B_R(x_0)} |Dv|^2 \, dx \, .$$

Consequently, we have for all $r \in (0, R/2)$

$$\int_{B_r(x_0)} |Dv|^2 \, dx \le c(n) r^n \sup_{B_{R/2}(x_0)} |Dv|^2$$

$$\le c(n,N,L) \left(\frac{r}{R}\right)^n \int_{B_R(x_0)} |Dv|^2 \, dx \, ,$$

and since the corresponding estimate for $r \in [R/2, R]$ is trivially satisfied with constant 2^n, we have also shown the Morrey type decay estimate (5.34) in the case $\varepsilon \in (n-2, n)$. Thus, the proof of the lemma is complete. □

Remark 5.29 This lemma uncovers a peculiarity of the two-dimensional case $n = 2$: the solution to the comparison problem (5.33) has Hölder continuous first derivatives, hence, we have $\text{Sing}_0(Dv) = \emptyset$. We will see later, that the regularity of Dv can be carried over to the solution of the original problem, which, in the two-dimensional case, is also everywhere of class C^1, see Theorem 5.31.

Morrey estimates in low dimensions up to $n = 4$ Once the comparison estimates are available, we can proceed with the proof of the partial regularity result for weak solutions, which is based on Morrey estimates for Du and the fact that whenever a $W^{1,2}$-function has its gradient in a Morrey space $L^{2,\mu}$ with $\mu > n - 2$, then it is automatically (Hölder) continuous. The latter Morrey condition can be shown to be satisfied in the low dimensional case $n \leq 4$. This approach to partial regularity traces its origins back to Campanato [11], who already observed that in $n = 2$ dimensions the partial continuity of weak solutions can be improved to everywhere continuity (and, under stronger assumptions on the vector field a, in fact to C^1-regularity, see Theorem 5.31).

Theorem 5.30 (Campanato)   *Consider* $n \in \{2, 3, 4\}$ *and let* $u \in W^{1,2}(\Omega, \mathbb{R}^N)$ *be a weak solution to the system* (5.1) *with a vector field* $a \colon \Omega \times \mathbb{R}^N \times \mathbb{R}^{Nn} \to \mathbb{R}^{Nn}$ *which satisfies the assumptions* (H1), (H2), *and* (H3'). *Then we have the characterization of the singular set via*

$$\mathrm{Sing}_0(u) = \left\{ x_0 \in \Omega \colon \liminf_{\varrho \searrow 0} \fint_{\Omega(x_0, \varrho)} |u - (u)_{\Omega(x_0, \varrho)}|^2 \, dx > 0 \right\},$$

with

$$\mathrm{Sing}_0(u) = \emptyset \qquad \text{for } n = 2$$

and

$$\dim_{\mathcal{H}}(\mathrm{Sing}_0(u)) < n - 2 \qquad \text{for } n \in \{3, 4\}.$$

Moreover, there exists $\delta > 0$ *depending only on* n *and* L *such that for every* $\lambda \in (0, \min\{1, 1 - (n - 2)/2 + \delta\})$ *we have* $\mathrm{Reg}_0(u) = \mathrm{Reg}_\lambda(u)$, *i.e.* $u \in C^{0,\lambda}(\mathrm{Reg}_0(u), \mathbb{R}^N)$.

Proof The proof is divided into a number of steps.

Definition of a comparison function. We start by fixing a ball $B_{2R}(x_0) \Subset \Omega$ and define $v \in u + W_0^{1,2}(B_R(x_0), \mathbb{R}^N)$ as the unique solution to the frozen system

$$\mathrm{div}\, a_0(Dv) = 0 \qquad \text{in } B_R(x_0),$$

where $a_0 \colon \mathbb{R}^{Nn} \to \mathbb{R}^{Nn}$ is defined as $a_0(z) := a(x_0, (u)_{B_R(x_0)}, z)$ for all $z \in \mathbb{R}^{Nn}$. The existence of the function v is guaranteed by the theory of monotone operators, see Theorem A.12 and Remark 5.27. Moreover, v is known to satisfy Morrey-type decay estimates, as stated in Lemma 5.28, which are going to be carried over to the weak solution u in the subsequent steps.

Energy control for Dv. We first show that the Dirichlet energy of v is controlled on the ball $B_R(x_0)$ by the Dirichlet energy of u. This follows easily by testing the frozen system with $u - v \in W_0^{1,2}(B_R(x_0), \mathbb{R}^N)$ and by taking into account the ellipticity estimate from Remark 5.1 (iii) combined with the growth assumption (H1). In this way, we find

$$\int_{B_R(x_0)} |Dv|^2 \, dx \leq \int_{B_R(x_0)} \big(a_0(Dv) - a_0(0)\big) \cdot Dv \, dx$$

$$= \int_{B_R(x_0)} \big(a_0(Dv) - a_0(0)\big) \cdot Du \, dx$$

$$\leq L \int_{B_R(x_0)} |Dv||Du| \, dx \, ,$$

and Hölder's inequality yields the desired energy control

$$\int_{B_R(x_0)} |Dv|^2 \, dx \leq L^2 \int_{B_R(x_0)} |Du|^2 \, dx \, .$$

Comparison estimate. Our next aim is to establish a comparison estimate between Du and Dv on $B_R(x_0)$. To this end, we take $u - v \in W_0^{1,2}(B_R(x_0), \mathbb{R}^N)$ as test functions for both the weak formulations of the frozen and the original system. With the ellipticity estimate from Remark 5.1 (iii) and the continuity assumption (H3'), we deduce

$$\int_{B_R(x_0)} |Du - Dv|^2 \, dx$$

$$\leq \int_{B_R(x_0)} \big(a_0(Du) - a_0(Dv)\big) \cdot (Du - Dv) \, dx$$

$$= \int_{B_R(x_0)} \big(a(x_0, (u)_{B_R(x_0)}, Du) - a(x, u, Du)\big) \cdot (Du - Dv) \, dx$$

$$\leq 2L \int_{B_R(x_0)} \omega\big(|x - x_0| + |u - (u)_{B_R(x_0)}|\big)(1 + |Du|)|Du - Dv| \, dx \quad (5.37)$$

(which should be compared to the intermediate estimate (4.28) in the direct approach). At this stage we recall the higher integrability of Du from Lemma 5.3, i.e., there exists an exponent $p > 2$ depending only on n, N and L such that $Du \in W^{1,p}(B_R(x_0), \mathbb{R}^{Nn})$ with the reverse Hölder-type estimate (5.3). Hence, we may apply Hölder's inequality with exponents $2p/(p-2), p$ and 2. In view of the assumption $\omega \leq 1$ we hence infer from (5.37)

$$\fint_{B_R(x_0)} |Du - Dv|^2 \, dx$$

$$\leq 4L^2 \left(\fint_{B_R(x_0)} \omega\big(R + |u - (u)_{B_R(x_0)}|\big) \, dx \right)^{1-\frac{2}{p}} \left(\fint_{B_R(x_0)} (1 + |Du|)^p \, dx \right)^{\frac{2}{p}}$$

$$\leq c \left(\fint_{B_R(x_0)} \omega\big(R + |u - (u)_{B_R(x_0)}|\big) \, dx \right)^{1-\frac{2}{p}} \fint_{B_{2R}(x_0)} (1 + |Du|^2) \, dx$$

with c depending only on n, N, and L. Now it only remains to bound the first integral on the right-hand side involving the modulus of continuity ω. Since ω is concave and satisfies $\omega(ct) \leq c\omega(t)$ for all $t \geq 0$ and $c \geq 1$, we find via Jensen's and Poincaré's inequality

$$\fint_{B_R(x_0)} \omega\big(R + |u - (u)_{B_R(x_0)}|\big) \, dx$$

$$\leq \omega \left(\fint_{B_R(x_0)} \big(R + |u - (u)_{B_R(x_0)}|\big) \, dx \right)$$

$$\leq \omega \left(\left(\fint_{B_R(x_0)} \big(R + |u - (u)_{B_R(x_0)}|\big)^2 \, dx \right)^{\frac{1}{2}} \right)$$

$$\leq c(n, N) \omega \left(\left(R^{2-n} \int_{B_R(x_0)} (1 + |Du|^2) \, dx \right)^{\frac{1}{2}} \right).$$

Combined with the previous estimate, we therefore end up with

$$\int_{B_R(x_0)} |Du - Dv|^2 \, dx$$

$$\leq c\omega^{1-\frac{2}{p}} \left(\left(R^{2-n} \int_{B_R(x_0)} (1 + |Du|^2) \, dx \right)^{\frac{1}{2}} \right) \int_{B_{2R}(x_0)} (1 + |Du|^2) \, dx.$$

with a constant c depending only on n, N, and L.

Morrey-type decay estimates for Du. With the help of the previous estimate we now carry the Morrey-type decay estimate for Dv from Lemma 5.28 over to Du. To this end, we essentially need to determine the decay of the quantity $\|Du\|_{L^2(B_\varrho(x_0), \mathbf{R}^{Nn})}$ with respect to $\varrho \in (0, R]$. We now take advantage of (5.34), the comparison and the energy estimate. This yields

$$\int_{B_\varrho(x_0)} \left(1 + |Du|^2\right) dx$$

$$\leq 2 \int_{B_\varrho(x_0)} \left(1 + |Dv|^2\right) dx + 2 \int_{B_\varrho(x_0)} |Du - Dv|^2 \, dx$$

$$\leq c \left(\frac{\varrho}{R}\right)^{\min\{n, 2+\varepsilon\}} \int_{B_R(x_0)} \left(1 + |Dv|^2\right) dx + 2 \int_{B_R(x_0)} |Du - Dv|^2 \, dx$$

$$\leq c \left[\left(\frac{\varrho}{R}\right)^{\min\{n, 2+\varepsilon\}} + \omega^{1-\frac{2}{p}} \left(\left(R^{2-n} \int_{B_R(x_0)} \left(1 + |Du|^2\right) dx\right)^{\frac{1}{2}}\right)\right]$$

$$\times \int_{B_{2R}(x_0)} \left(1 + |Du|^2\right) dx \,,$$

with a constant c depending only n, N, and L, and with a positive number $\varepsilon > 0$ depending only on n and L (in particular independent of x_0). This inequality extends easily to all $\varrho \in (0, 2R]$ and will now be the clue to the partial regularity result.

Characterization and regularity improvement. We start by observing the identity

$$\left\{x_0 \in \Omega \colon \liminf_{\varrho \searrow 0} \fint_{\Omega(x_0, \varrho)} |u - (u)_{\Omega(x_0, \varrho)}|^2 \, dx > 0\right\}$$

$$= \left\{x_0 \in \Omega \colon \liminf_{\varrho \searrow 0} \varrho^{2-n} \int_{\Omega(x_0, \varrho)} |Du|^2 \, dx > 0\right\},$$

which, similarly to (4.24), follows from Poincaré's inequality and the Caccioppoli inequality in Proposition 5.2 (with the choice $\zeta = (u)_{B_{2\varrho}(x_0)}$) via

$$\fint_{B_\varrho(x_0)} |u - (u)_{B_\varrho(x_0)}|^2 \, dx \leq c(n, N) \varrho^{2-n} \int_{B_\varrho(x_0)} |Du|^2 \, dx$$

$$\leq c(n, N, L) \left(\varrho^2 + \fint_{B_{2\varrho}(x_0)} |u - (u)_{B_{2\varrho}(x_0)}|^2 \, dx\right)$$

for each ball $B_{2\varrho}(x_0) \subset \Omega$. Therefore, we now assume that $x_0 \in \Omega$ satisfies

$$\liminf_{\varrho \searrow 0} \varrho^{2-n} \int_{\Omega(x_0, \varrho)} |Du|^2 \, dx = 0 \,,$$

and we need to show $x_0 \in \text{Reg}_\lambda(u)$ (the reverse implication is again obvious). This choice implies that the factor

$$\omega^{1-\frac{2}{p}}\left(\left(R^{2-n}\int_{B_R(x_0)} (1+|Du|^2)\,dx\right)^{\frac{1}{2}}\right)$$

appearing in the final inequality of the previous step can be made smaller than any given positive number κ provided that $R < R_0$ is sufficiently small (and as in the proof of Theorem 5.9, such a pointwise smallness condition is in fact satisfied in a small neighbourhood of x_0). Hence, as a consequence of Lemma B.3, we obtain a Morrey-type estimate of the form

$$\int_{B_\varrho(x_0)} (1+|Du|^2)\,dx \le \varrho^{\min\{n,2+\varepsilon\}-\varepsilon'} \tag{5.38}$$

for every $\varepsilon' \in (0, \min\{n, 2+\varepsilon\})$ whenever ϱ is sufficiently small in dependence of n, N, L, ε', and x_0. Indeed, since the dependence on the point x_0 is continuous, we deduce $Du \in L^{2,\min\{n,2+\varepsilon\}-\varepsilon'}$ locally in a neighbourhood of x_0.

Only at this point, the low dimensional assumption $n \le 4$ enters and we observe that the exponent $\min\{n, 2 + \varepsilon\}$ is (strictly) bounded from below by $n - 2$, which is crucial for the application of Corollary 1.58. For $n = 2$, we obtain immediately continuity in a neighbourhood of x_0 with any Hölder exponent in $(0,1)$ (since ε' is arbitrary). Otherwise, for $n \in \{3, 4\}$, we define

$$\delta := \frac{\min\{n, \varepsilon + 2\} - \varepsilon' - 2}{2}$$

which is strictly positive, if we restrict ourselves to $\varepsilon' < \min\{1, \varepsilon\}$. With this choice, the asserted Hölder continuity with any exponent $\lambda \in (0, \min\{1, 1 - (n-2)/2 + \delta\})$ follows again from Corollary 1.58 (in the localized version of Remark 1.60), and the proof of the regularity improvement is complete.

Hausdorff dimension of $\text{Sing}_0(u)$. We again recall the higher integrability result for Du from Lemma 5.3, for some integrability exponent $p > 2$ depending only on n, N, and L. In the case $n = 2$ Morrey's inequality from Theorem 1.61 implies $\text{Sing}_0(u) = \emptyset$, while for $n > 2$, we can improve the condition of x_0 being a regular point via

$$\varrho^{p-n}\int_{B_\varrho(x_0)} (1+|Du|^p)\,dx \le c(n, N, L)\left(\varrho^{2-n}\int_{B_{2\varrho}(x_0)} (1+|Du|^2)\,dx\right)^{\frac{p}{2}}$$

for every ball $B_{2\varrho}(x_0) \subset \Omega$. Consequently, we get

$$\text{Sing}_0(u) \subset \left\{x_0 \in \Omega: \liminf_{\varrho \searrow 0} \varrho^{p-n}\int_{\Omega(x_0,\varrho)} (1+|Du|^p)\,dx > 0\right\}$$

which, in view of Lemma 1.74 (applied with the measure of a set defined via integration of the function $1 + |Du|^p$ over this set, cf. Remark 1.75), in turn provides the strict upper bound $n - 2$ for the Hausdorff dimension of the singular set.   □

Everywhere C^1-regularity in the two-dimensional case Finally, we provide a regularity improvement in dimensions $n = 2$, under the sole additional assumption that the vector field a is Hölder continuous with respect to the first two variables, and not only continuous. Similarly as for the proof of the previous Theorem 5.30, the underlying idea for the proof of the two-dimensional statement is a a direct comparison argument, but now Campanato-type (instead of Morrey-type) decay estimates are carried over from the solution of the frozen system. We here follow [52, Section 9], where the corresponding statement was inferred for minimizers of convex variational integrals, and [4].

Theorem 5.31 *Consider $n = 2$ and let $u \in W^{1,2}(\Omega, \mathbb{R}^N)$ be a weak solution to the system (5.1) with a vector field $a \colon \Omega \times \mathbb{R}^N \times \mathbb{R}^{2N} \to \mathbb{R}^{2N}$ satisfying the assumptions (H1), (H2), and (H3). Then $\mathrm{Sing}_0(Du) = \emptyset$ and $u \in C^{1,\alpha}(\Omega, \mathbb{R}^N)$.*

Proof We fix a ball $B_R(x_0) \Subset \Omega$ and define $v \in u + W_0^{1,2}(B_R(x_0), \mathbb{R}^N)$ as the unique solution to the frozen system

$$\mathrm{div}\, a_0(Dv) = 0 \qquad \text{in } B_R(x_0),$$

where $a_0 \colon \mathbb{R}^{2N} \to \mathbb{R}^{2N}$ is defined as $a_0(z) := a(x_0, (u)_{B_R(x_0)}, z)$ for all $z \in \mathbb{R}^{2N}$. According to Lemma 5.28, Dv is known to satisfy a Campanato-type decay estimate, which is now going to be carried over to u.

We first provide a refinement of the comparison estimates which was established in the proof of Theorem 5.30. Taking advantage of the Hölder continuity of the vector field a with respect to the (x, u)-variable with exponent α and of the solution u with any exponent $\lambda \in (0, 1)$ (provided by Theorem 5.30), we enter in the comparison estimate (5.37) and infer

$$\int_{B_R(x_0)} |Du - Dv|^2\, dx \leq c\big(L, [u]_{C^{0,\lambda}(B_R(x_0), \mathbb{R}^N)}\big) R^{\alpha\lambda} \int_{B_R(x_0)} \big(1 + |Du|^2\big)\, dx.$$

Involving also the Morrey-type decay estimate (5.38), we know that the integral on the right-hand side decays as $R^{2-\varepsilon'}$ for any $\varepsilon' \in (0, 2)$, if we restrict ourselves to sufficiently small values of $R < R_0$. We now take advantage of the decay estimate (5.35) for Dv, the refined comparison and the energy estimate. This yields for any $\varrho \leq R < R_0$:

$$\int_{B_\varrho(x_0)} |Du - (Du)_{B_\varrho(x_0)}|^2 \, dx$$

$$\leq 2 \int_{B_\varrho(x_0)} |Dv - (Dv)_{B_\varrho(x_0)}|^2 \, dx + 2 \int_{B_\varrho(x_0)} |Du - Dv|^2 \, dx$$

$$\leq c \Big(\frac{\varrho}{R}\Big)^{2+\varepsilon} \int_{B_R(x_0)} |Dv - (Dv)_{B_R(x_0)}|^2 \, dx + 2 \int_{B_R(x_0)} |Du - Dv|^2 \, dx$$

$$\leq c \Big[\Big(\frac{\varrho}{R}\Big)^{2+\varepsilon} + R^{\alpha\lambda}\Big] \int_{B_R(x_0)} \big(1 + |Du|^2\big) \, dx$$

$$\leq c \Big[\Big(\frac{\varrho}{R}\Big)^{2+\varepsilon} + R^{\alpha\lambda}\Big] R^{2-\varepsilon'},$$

with a constant c depending only n, N, and L, and with a positive number $\varepsilon > 0$. If $\varrho < R$ is related to R via

$$R = \varrho^{\frac{2+\varepsilon}{\alpha\lambda+2+\varepsilon}},$$

then the previous inequality reduces to the Campanato decay estimate

$$\int_{B_\varrho(x_0)} |Du - (Du)_{B_\varrho(x_0)}|^2 \, dx \leq c\varrho^{(2+\varepsilon)\frac{\alpha\lambda+2-\varepsilon'}{\alpha\lambda+2+\varepsilon}} = c\varrho^{2+\frac{\alpha\lambda\varepsilon-(2+\varepsilon)\varepsilon'}{\alpha\lambda+2+\varepsilon}}.$$

For ε' sufficiently small, the exponent on the right-hand side is greater than the space dimension 2. Since the same Campanato estimate is available for all points in a neighbourhood of x_0 by absolute continuity of the integral, we conclude from Theorem 1.27 that Du is in particular locally continuous in Ω. This proves $\text{Sing}_0(Du) = \emptyset$, and the optimal Hölder regularity of Du with exponent α then follows from the identity $\text{Reg}_0(Du) = \text{Reg}_\alpha(Du)$ established in Theorem 5.9. $\square$

Remarks 5.32 For convex variational integrals there are similar dimension reduction results available as those presented here for elliptic systems. Specifically, we mention that for every minimizer u of the functional F defined in (5.18) with integrand f satisfying the assumptions (F0), (F1), (F2), (5.19), and (F4), Kristensen and Mingione [52] have established the following (in fact, slightly better) bounds on the Hausdorff dimension of the singular set of Du (building on its characterization given in Theorem 5.12 and with a similar reasoning as presented above):

(i) If f does not depend explicitly on the u-variable, then there holds $\dim_{\mathcal{H}}(\text{Sing}_0(Du)) \leq n - \alpha$.

(ii) In dimensions $n \in \{3, 4\}$ there holds $\dim_{\mathcal{H}}(\text{Sing}_0(Du)) \leq n - \alpha$.

(iii) In the two-dimensional case $n = 2$ there holds $\mathrm{Sing}_0(Du) = \emptyset$.

(iv) In the general case there holds $\dim_{\mathcal{H}}(\mathrm{Sing}_0(Du)) \leq n - \min\{\alpha, p-2\}$, with $p > 2$ the higher integrability exponent of Du (hence, also here the partial regularity result from Theorem 5.12 is never optimal).

Appendix A
Functional Analysis

We here provide some basic tools (without proofs) from linear functional analysis, which are used in the course of these lecture notes. We first gather the structures of metric, norm and inner product on general spaces, which can be interpreted as natural generalizations of the respective notions on the euclidean space $\mathbb{R}^n$. Then we recall the definitions of dual spaces, weak convergence and weak compactness, and present some essential facts in this regard. Finally, we state the theorems of Lax–Milgram and of Browder–Minty, which are relevant for the existence theory for linear and nonlinear partial differential equations.

Metric spaces We start by defining the notion of a distance on general sets.

Definition A.1 Let X be a set. A map $d\colon X \times X \to \mathbb{R}_0^+$ is called *metric* if there hold

(i) $d(x,y) \le d(x,z) + d(z,y)$ for all $x, y, z \in X$ (triangle inequality),
(ii) $d(x,y) = d(y,x)$ for all $x, y \in X$ (symmetry),
(iii) $d(x,y) = 0$ if and only if $x = y$.

Moreover, the pair (X, d) of a set X and a metric d on X is called a *metric space* (but explicit reference to the metric is usually omitted if it is clear from the context which metric is used).

In a metric space it is possible to introduce topological objects like closed and open sets (e.g. the balls $B_r(x_0)$ in the euclidean space $\mathbb{R}^n$ with radius r). In a more abstract sense, such topological properties can be used to study *topological spaces* that require for its basic definition only a collection of subsets of the set X (the open sets) which satisfy the conditions of a topology. However, in what follows, we shall only need the concepts of Cauchy sequences, convergence of sequences, separability and completeness in metric spaces.

© Springer International Publishing Switzerland 2016
L. Beck, *Elliptic Regularity Theory*, Lecture Notes of the Unione
Matematica Italiana 19, DOI 10.1007/978-3-319-27485-0

Definition A.2 Let (X, d) be a metric space.

(i) A sequence $(x_j)_{j \in \mathbb{N}}$ in X is called a *Cauchy sequence* if there holds $d(x_j, x_\ell) \to 0$ as $j, \ell \to \infty$;

(ii) A sequence $(x_j)_{j \in \mathbb{N}}$ in X is said to *converge* to some $x \in X$ (written $x_j \to x$) if $d(x_j, x) \to 0$ as $j \to \infty$;

(iii) X is called *separable* if there exists a countable, dense subset;

(iv) X is called *complete* if every Cauchy sequence in X converges.

Normed spaces If the underlying space is not a general set but a vector space, one can define the notion of a norm.

Definition A.3 Let X be a vector space. A map $\| \cdot \|_X \colon X \to \mathbb{R}_0^+$ is called a *norm* if there hold

(i) $\|x + y\|_X \leq \|x\|_X + \|y\|_X$ for all $x, y \in X$ (triangle inequality),

(ii) $\|\lambda x\|_X = |\lambda| \|x\|_X$ for all $x \in X$ and $\lambda \in \mathbb{R}$ (homogeneity),

(iii) $\|x\|_X = 0$ implies $x = 0$.

Moreover, the pair $(X, \| \cdot \|_X)$ of a vector space X and a norm $\| \cdot \|_X$ on X is called a *normed space* (but explicit reference to the norm is usually omitted if it is clear from the context which norm is used).

By defining the map $d \colon X \times X \to \mathbb{R}_0^+$ via $d(x, y) := \|x - y\|_X$ for all $x, y \in X$, every norm induces a metric in a natural way. In this sense, one can define Cauchy sequences, convergence of sequences (also referred to as strong convergence, that is, convergence in norm), and completeness for normed spaces exactly as for metric spaces. We further call a normed space a *Banach space* if it is complete with respect to the metric induced by the norm.

Examples of Banach spaces, which appear in these notes, are Hölder spaces $C^{k,\alpha}(\overline{\Omega})$, Lebesgue spaces $L^p(\Omega)$, Morrey spaces $L^{p,\lambda}(\Omega)$, Campanato spaces $\mathcal{L}^{p,\lambda}(\Omega)$, and Sobolev spaces $W^{k,p}(\Omega)$, for $k \in \mathbb{N}$, $p \in [1, \infty]$, $\lambda \in [0, \infty)$ and $\alpha \in [0, 1]$, cp. Theorems 1.4, 1.18 and 1.33.

Definition A.4 Let X, Y be two Banach spaces with $X \subset Y$. We say that X is *compactly embedded* in Y (written $X \Subset Y$) provided that

(i) X is continuously embedded in Y, i.e., there exists a constant C such that $\|x\|_Y \leq C \|x\|_X$ for all $x \in X$;

(ii) Every bounded sequence in X is sequentially precompact in Y, i.e., it has a subsequence which converges in Y.

Remark A.5 If X, Y, Z are Banach spaces with $X \subset Y \subset Z$ and if one of the embeddings $X \hookrightarrow Y$ and $Y \hookrightarrow Z$ is compact and the other one continuous, then the composite embedding $X \hookrightarrow Z$ is also compact.

Definition A.6 Let X be a real vector space. A map $\langle \cdot, \cdot \rangle_X \colon X \times X \to \mathbb{R}$ is called an *inner product* if there hold

(i) $\langle x, y \rangle_X = \langle y, x \rangle_X$ for all $x, y \in X$ (symmetry),

(ii) the map $x \mapsto \langle x, y \rangle_X$ is linear for each $y \in X$ (linearity),

(iii) $\langle x, x \rangle_X \geq 0$ for all $x \in X$ and $\langle x, x \rangle_X = 0$ if and only if $x = 0$ (positive definiteness).

Given a real vector space X with inner product $\langle \cdot, \cdot \rangle_X$, one can define a norm $\| \cdot \|_X$ via $\|x\|_X := \sqrt{\langle x, x \rangle_X}$ for all $x \in X$. For the verification of the triangle inequality, we mention the Cauchy–Schwarz inequality $|\langle x, y \rangle_X| \leq \|x\|_X \|y\|_X$ for all $x, y, \in X$, which is a direct consequence of the positivity of $\langle x + \lambda y, x + \lambda y \rangle_X$ for a suitable choice of λ. Finally, we call a vector space *Hilbert space* if it is a Banach space with the norm induced by the inner product.

Examples of Hilbert spaces are the Lebesgue space $L^2(\Omega)$ and the Sobolev space $W^{1,2}(\Omega)$, cp. Remarks 1.20 and 1.34.

Linear maps on normed vector spaces, dual spaces and weak convergence We here restrict ourselves to normed spaces, even though parts of the theory could be developed in topological spaces as well. So let X and Y be two normed spaces, with norms indicated by $\| \cdot \|_X$ and $\| \cdot \|_Y$, respectively. We define

$$L(X;Y) := \{T : X \to Y : T \text{ is linear and continuous}\}$$

(and note that continuity and boundedness are equivalent for a linear map $T : X \to Y$). We further define $\| \cdot \|_{L(X;Y)} : L(X;Y) \to \mathbb{R}_0^+$ by setting for each element $T \in L(X;Y)$

$$\|T\|_{L(X;Y)} := \sup_{x \in X : \|x\|_X \leq 1} \|Tx\|_Y .$$

This is indeed a norm on $L(X;Y)$ (called the operator norm), and endowed with it the space $L(X;Y)$ is a normed space. Furthermore, if Y is a Banach space, then so is $L(X;Y)$ (and the reverse implication is also true whenever $X \neq \{0\}$ is non-trivial). We are particularly interested in the case $Y = \mathbb{R}$.

Definition A.7 Let X be a normed space.

(i) The space $X^* := L(X; \mathbb{R})$ is called the *dual space* of X, and its elements are called *bounded linear functionals* on X. The space $X^{**} := (X^*)^*$ is further called the *bidual space* of X;

(ii) The *duality pairing* $\langle \cdot, \cdot \rangle_{X, X^*} : X \times X^* \to \mathbb{R}$ of X and X^* is defined via

$$\langle x, x^* \rangle_{X, X^*} := x^*(x)$$

for all $x \in X$ and $x^* \in X^*$;

(iii) A Banach space X is called *reflexive* if the map $J_X \in L(X; X^{**})$, defined via $J_X(x)(x^*) := \langle x, x^* \rangle_{X, X^*}$ for all $x \in X$ and $x^* \in X^*$, is surjective, i.e., if for every $x^{**} \in X^{**}$ there exists some $x \in X$ such that

$$\langle x^*, x^{**} \rangle_{X^*, X^{**}} = \langle x, x^* \rangle_{X, X^*} \qquad \text{for all } x^* \in X^*.$$

Examples of reflexive spaces are the Lebesgue spaces $L^p(\Omega)$ and the Sobolev spaces $W^{k,p}(\Omega)$ for $k \in \mathbb{N}$ and $p \in (1, \infty)$.

With the dual space at hand, we can now define weaker notions of convergence and compactness.

Definition A.8 Let X be a Banach space.

(i) A sequence $(x_j)_{j \in \mathbb{N}}$ in X is said to *converge weakly* to some $x \in X$ (written $x_j \rightharpoonup x$) if for every $x^* \in X^*$ there holds $\langle x_j, x^* \rangle_{X, X^*} \to \langle x, x^* \rangle_{X, X^*}$ as $j \to \infty$;

(ii) A sequence $(x_j^*)_{j \in \mathbb{N}}$ in X^* is said to *converge weakly-$*$* to some $x^* \in X^*$ (written $x_j^* \overset{*}{\rightharpoonup} x^*$) if for every $x \in X$ there holds $\langle x, x_j^* \rangle_{X, X^*} \to \langle x, x^* \rangle_{X, X^*}$ as $j \to \infty$;

(iii) A set $C \subset X$ is called *weakly (sequentially) closed* if every weakly convergent subsequence has its weak limit again in C;

(iv) A set $K \subset X$ is called *weakly (sequentially) compact* if every sequence in K has a weakly convergent subsequence, with weak limit again in K;

(v) A set $K^* \subset X^*$ is called *weakly-$*$ (sequentially) compact* if every sequence in K^* has a weakly-$*$ convergent subsequence, with weak-$*$ limit again in K^*.

Even though it is not needed here, we note that one can define a notion of weak compactness also via coverings. To do so, one defines the *weak topology* as the coarsest topology on X such that every $x^* \in X^*$ is continuous (analogously, the weak-$*$ topology is the coarsest topology on X^* such that $J_X(x)$ is continuous for every $x \in X$). One then defines a set $K \subset X$ to be *weakly compact* (or weakly-$*$ compact) if every open cover of K with sets in the weak topology (or weak-$*$ topology) admits a finite subcover. Unless X is a metric space, these concepts of compactness are in general different. Concerning closedness, we note that every set which is weakly closed is also (strongly) closed, but the converse is in general false. However, it is true for convex sets (known as Mazur's lemma), and this is of some relevance for partial differential equations (in particular the existence theory).

Lemma A.9 (Mazur) *Let X be a Banach space. A convex subset of X is closed if and only if it is weakly closed.*

We now comment on some basic facts on weak and weak-$*$-convergence. We first observe that convergence in norm (that is, strong convergence) obviously implies weak (and weak-$*$) convergence. Similarly as for convergence in norm, we also have uniqueness of weak limits (as a consequence of Hahn–Banach theorem), whereas uniqueness of weak-$*$-limits is trivially true. Finally, we mention that the norm is lower semicontinuous with respect to weak and weak-∗-convergence (the proof of the first statement again involves a Hahn–Banach-type argument).

Via the Banach–Steinhaus theorem one can show that weak (and also weak-∗) convergent sequences are bounded. Conversely, we can extract a weakly convergent subsequence from an arbitrary bounded sequence if and only if the underlying space is reflexive. More precisely, we have the following equivalences.

Theorem A.10 *Let X be a Banach space. Then the following statements are equivalent:*

(i) *X is reflexive;*
(ii) *The dual space X^* is reflexive;*
(iii) *The closed unit ball $B := \{x \in X \colon \|x\|_X \leq 1\}$ is weakly compact;*
(iv) *Every bounded sequence in X has a weakly convergent subsequence.*

In the literature, these statement can for example be found in [6]. The equivalence of (i) and (ii) is stated in [6, Corollary 3.21], the equivalence of (i) and (iii) in [6, Theorem 3.17] (due to Kakutani), and the equivalence of (i) and (iv) in [6, Theorem 3.18 and Theorem 3.19] (due to Eberlein and Šmulian).

Some statements of relevance for the existence theory The first statement in this regard concerns the Lax–Milgram theorem which can be seen as a generalization of the Riesz representation theorem and which in particular can be used to show the existence and uniqueness of a weak solution to the Dirichlet problem for linear elliptic systems (see Remark 4.12).

Theorem A.11 (Lax–Milgram) *Let H be a real Hilbert space and let $B \colon H \times H \to \mathbb{R}$ be a bilinear form which is bounded and coercive, i.e., there exists a constant $L \geq 1$ such that*

$$B(v, w) \leq L\|v\|_H \|w\|_H \,,$$
$$B(v, v) \geq \|v\|_H^2$$

hold for all $v, w \in H$. Then there exists a linear bijection $\Lambda \colon H^* \to H$ such that

$$B(\Lambda(F), v) = F(v)$$

for all $F \in H^*$ and all $v \in H$. Moreover, both Λ and its inverse Λ^{-1} are bounded.

For the proof we refer to [84, Chapter III.7]. Secondly, we recall a basic existence and uniqueness result from the theory of monotone operators, cf. [56, Théorème 2.2.1], which is sufficiently general to provide the existence of weak solutions also to the Dirichlet problem for some nonlinear elliptic systems (see Remark 5.27).

Theorem A.12 (Browder–Minty)   *Let X be a reflexive, separable Banach space and let $A\colon X \to X^*$ be a bounded, continuous operator which is (strictly) monotone and coercive, i.e., there hold*

$$\langle v - w, A(v) - A(w)\rangle_{X,X^*} \geq (>)\,0 \quad \text{for all } v \neq w \in X,$$

$$\langle v, A(v)\rangle_{X,X^*}\|v\|_X^{-1} \to \infty \quad \text{as } \|v\|_X \to \infty.$$

Then the operator A is (injective and) surjective, i.e. for every $F \in X^*$ there exists (a unique) $u \in X$ such that $A(u) = F$.

Appendix B
Some Technical Lemmata

We here collect some (well-known) iteration lemmata.

Lemma B.1 ([30], Lemma V.3.1) *Assume that $\phi(\varrho)$ is a non-negative, real-valued, bounded function defined on an interval $[r, R] \subset \mathbb{R}^+$. Assume further that for all $r \leq \varrho < \sigma \leq R$ we have*

$$\phi(\varrho) \leq \left[A_1(\sigma - \varrho)^{-\alpha_1} + A_2(\sigma - \varrho)^{-\alpha_2} + A_3\right] + \vartheta\phi(\sigma)$$

for some non-negative constants A_1, A_2, A_3, non-negative exponents $\alpha_1 \geq \alpha_2$, and a parameter $\vartheta \in [0, 1)$. Then we have

$$\phi(r) \leq c(\alpha_1, \vartheta)\left[A_1(R - r)^{-\alpha_1} + A_2(R - r)^{-\alpha_2} + A_3\right].$$

Proof We proceed by iteration and start by defining a sequence $(\varrho_j)_{j \in \mathbb{N}_0}$ via

$$\varrho_j := r + (1 - \lambda^j)(R - r)$$

for some $\lambda \in (0, 1)$ to be chosen later. This sequence is increasing, converging to R as $j \to \infty$, and the difference of two subsequent members is given by

$$\varrho_j - \varrho_{j-1} = (1 - \lambda)\lambda^{j-1}(R - r).$$

Applying the assumption inductively with $\varrho = \varrho_{j-1}$, $\sigma = \varrho_j$ for $j \in \{1, \ldots, \ell\}$ and taking into account $\alpha_1 \geq \alpha_2$, we obtain

$$\phi(r) = \phi(\varrho_0)$$

$$\leq A_1(1 - \lambda)^{-\alpha_1}(R - r)^{-\alpha_1} + A_2(1 - \lambda)^{-\alpha_2}(R - r)^{-\alpha_2} + A_3 + \vartheta\phi(\varrho_1)$$

$$\leq (1 - \lambda)^{-\alpha_1}\sum_{j=0}^{\ell-1}\vartheta^j\lambda^{-j\alpha_1}\left[A_1(R - r)^{-\alpha_1} + A_2(R - r)^{-\alpha_2} + A_3\right] + \vartheta^\ell\phi(\varrho_\ell)$$

© Springer International Publishing Switzerland 2016
L. Beck, *Elliptic Regularity Theory*, Lecture Notes of the Unione
Matematica Italiana 19, DOI 10.1007/978-3-319-27485-0

for every $\ell \in \mathbb{N}$. If we now choose λ in dependency of ϑ and α_1 such that $\vartheta \lambda^{-\alpha_1} < 1$, then the series on the right-hand side of the previous inequality converges. Therefore, passing to the limit $\ell \to \infty$, we arrive at the conclusion, with constant $c(\alpha_1, \vartheta) := (1 - \lambda)^{-\alpha_1}(1 - \vartheta \lambda^{-\alpha_1})^{-1}$.   □

Lemma B.2 ([78], Lemma 5.1)   *Assume that $\phi(h, \varrho)$ is a non-negative, real-valued function defined for $h \geq k_0$ and $r \leq \varrho \leq R$. Assume further that it is non-increasing in h for fixed ϱ, that it is non-decreasing in ϱ for fixed h, and that for all $k > h > k_0$ and $r \leq \varrho < \sigma \leq R$ we have*

$$\phi(k, \varrho) \leq \left[A_1 (k - h)^{-\alpha_1}(\sigma - \varrho)^{-\alpha_2} + A_2 (k - h)^{-\alpha_1 - \alpha_2} \right] \phi(h, \sigma)^\beta \quad \text{(B.1)}$$

with constants $A_1 > 0$, $A_2 \geq 0$, with positive exponents α_1, α_2, and a parameter $\beta > 1$. Then we have

$$\phi(k_0 + d, r) = 0$$

where d is given by

$$d^{\alpha_1} = A_1 (R - r)^{-\alpha_2} 2^{\frac{\beta(\alpha_1 + \alpha_2)}{\beta - 1} + 1} \phi(k_0, R)^{\beta - 1} + (A_1^{-1} A_2)^{\frac{\alpha_1}{\alpha_2}} (R - r)^{\alpha_1}.$$

Proof We proceed by iteration, and for this purpose, we define two sequences $(k_j)_{j \in \mathbb{N}_0}$ and $(\varrho_j)_{j \in \mathbb{N}_0}$ via

$$k_j := k_0 + d(1 - 2^{-j}) \quad \text{and} \quad \varrho_j := r + 2^{-j}(R - r).$$

We observe that the sequence $(k_j)_{j \in \mathbb{N}_0}$ is increasing with limit $k_0 + d$, whereas the sequence $(\varrho_j)_{j \in \mathbb{N}_0}$ is decreasing with limit r. Furthermore, the differences of two subsequent members are given by

$$k_j - k_{j-1} = 2^{-j} d \quad \text{and} \quad \varrho_{j-1} - \varrho_j = 2^{-j}(R - r).$$

Applying the assumption of the lemma with $k = k_j$, $h = k_{j-1}$ and $\varrho = \varrho_j$, $\sigma = \varrho_{j-1}$ for arbitrary $j \in \mathbb{N}$, we obtain

$$\phi(k_j, \varrho_j) \leq \left[A_1 d^{-\alpha_1}(R - r)^{-\alpha_2} + A_2 d^{-\alpha_1 - \alpha_2} \right] 2^{(\alpha_1 + \alpha_2)j} \phi(k_{j-1}, \varrho_{j-1})^\beta$$

$$\leq A_1 d^{-\alpha_1}(R - r)^{-\alpha_2} 2^{1 + (\alpha_1 + \alpha_2)j} \phi(k_{j-1}, \varrho_{j-1})^\beta,$$

where we have also used $d^{\alpha_2} \geq A_1^{-1} A_2 (R - r)^{\alpha_2}$, available from the definition of d. In the next step we prove by induction that the estimate

$$\phi(k_j, \varrho_j) \leq 2^{-j \frac{\alpha_1 + \alpha_2}{\beta - 1}} \phi(k_0, \varrho_0) \quad \text{(B.2)}$$

holds for all $j \in \mathbb{N}_0$. Indeed, this inequality is trivial for $j = 0$. For the inductive step with $j \in \mathbb{N}$ we employ the definition of d as given in the statement (and note $\varrho_0 = R$) in order to find:

$$\phi(k_j, \varrho_j) \leq A_1 d^{-\alpha_1} (R - r)^{-\alpha_2} 2^{1+(\alpha_1+\alpha_2)j} \phi(k_{j-1}, \varrho_{j-1})^\beta$$

$$\leq A_1 d^{-\alpha_1} (R - r)^{-\alpha_2} 2^{1+(\alpha_1+\alpha_2)j} 2^{-(j-1)\beta \frac{\alpha_1+\alpha_2}{\beta-1}} \phi(k_0, \varrho_0)^{\beta-1} \phi(k_0, \varrho_0)$$

$$= 2^{-\beta \frac{\alpha_1+\alpha_2}{\beta-1} + (\alpha_1+\alpha_2)j - (j-1)\beta \frac{\alpha_1+\alpha_2}{\beta-1}} \phi(k_0, \varrho_0) = 2^{-j \frac{\alpha_1+\alpha_2}{\beta-1}} \phi(k_0, \varrho_0).$$

By the monotonicity properties of $\phi(h, \varrho)$ we deduce from (B.2)

$$\phi(k_0 + d, r) \leq \phi(k_j, \varrho_j) \leq 2^{-j \frac{\alpha_1+\alpha_2}{\beta-1}} \phi(k_0, R),$$

and the assertion follows from the passage to the limit $j \to \infty$.   □

Lemma B.3 *Assume that $\phi(\varrho)$ is a non-negative, real-valued, non-decreasing function defined on the interval $[0, R_0]$. Assume further that there exists a number $\tau \in (0, 1)$ such that for all $R \leq R_0$ we have*

$$\phi(\tau R) \leq (\tau^{\alpha_1} + \kappa)\phi(R) + AR^{\alpha_2}$$

for some non-negative constant A, some number $\kappa \geq 0$, and positive exponents $\alpha_1 > \alpha_2$. Then there exists a positive number $\kappa_0 = \kappa_0(\tau, \alpha_1, \alpha_2)$ such that for $\kappa \leq \kappa_0$ and all $r \leq R \leq R_0$ we have

$$\phi(r) \leq c(\tau, \alpha_1, \alpha_2)\left[\left(\frac{r}{R}\right)^{\alpha_2} \phi(R) + Ar^{\alpha_2}\right].$$

Proof We start by fixing some exponent $\alpha_3 \in (\alpha_2, \alpha_1)$ depending only on α_1 and α_2. We then determine $\kappa_0 \in (0, 1)$ such that $\tau^{\alpha_1} + \kappa_0 = \tau^{\alpha_3}$. By induction, we obtain for every $R \leq R_0$ and every $j \in \mathbb{N}$

$$\phi(\tau^j R) \leq \tau^{\alpha_3} \phi(\tau^{j-1} R) + A\tau^{(j-1)\alpha_2} R^{\alpha_2}$$

$$\leq \tau^{j\alpha_3} \phi(R) + A\tau^{(j-1)\alpha_2} R^{\alpha_2} \sum_{i=0}^{j-1} \tau^{i(\alpha_3-\alpha_2)}$$

$$\leq \tau^{j\alpha_2} \phi(R) + c(\tau, \alpha_1, \alpha_2) A\tau^{j\alpha_2} R^{\alpha_2}$$

where we have used the fact that the series is convergent. For an arbitrary $r \in (0, R]$ we determine $j \in \mathbb{N}_0$ such that $\tau^{j+1} R < r \leq \tau^j R$. Since ϕ is non-decreasing, the assertion follows from the computation

$$\phi(r) \leq \phi(\tau^j R) \leq \tau^{j\alpha_2}\phi(R) + c(\tau,\alpha_1,\alpha_2)A\tau^{j\alpha_2}R^{\alpha_2}$$
$$\leq \tau^{-\alpha_2}\left[\left(\frac{r}{R}\right)^{\alpha_2}\phi(R) + c(\tau,\alpha_1,\alpha_2)Ar^{\alpha_2}\right]. \qquad \Box$$

List of Notation

We here summarize most of the notation used in these lecture notes. The page number refers to the first occurrence or of the precise definition for each notation or abbreviation.

Sets and measures

$\emptyset$	empty set, p. 44
$A(k, x_0, r)$	k-super-level set of the relevant function in the ball $B_r(x_0)$, p. 61
$B(k, x_0, r)$	k-sub-level set of the relevant function in the ball $B_r(x_0)$, p. 61
$B_r(x_0)$	open n-dimensional ball with radius r and center x_0, p. 6
B_r	abbreviation for $B_r(0)$, p. 25
$\operatorname{diam}(S)$	diameter of the set S, p. 44
$\dim_{\mathcal{H}}(S)$	Hausdorff dimension of the set S, p. 45
$\operatorname{dist}(S, T)$	distance between two sets S and T, p. 27
$\mathcal{H}^k(S)$	k-dimensional Hausdorff measure of the set S, p. 44
$\mathcal{L}^n(S)$ or $|S|$	Lebesgue measure of the set S, p. 6
$\mathbb{N}$	the set of positive integers (and $\mathbb{N}_0 := \mathbb{N} \cup \{0\}$), p. 2
Ω	denotes usually a set in $\mathbb{R}^n$, on which a function or equation is given, p. 2
$\Omega(x_0, r)$	abbreviation for $B_r(x_0) \cap \Omega$, p. 11
ω_k	abbreviation for $\pi^{k/2}/\Gamma(1 + k/2)$, p. 44
$\mathbb{R}^n$	n-dimensional real Euclidean space, p. 2
$\mathbb{R}$	the real line (and $\mathbb{R}^+ := (0, \infty)$ and $\mathbb{R}_0^+ := \mathbb{R}^+ \cup \{0\}$), p. 2
$S \subset T$	the set S is (not necessarily strictly) contained in the set T, p. 3
$S \Subset T$	the set S is compactly contained in the set T, p. 4
$\overline{S}$	closure of the set S, p. 2
∂S	(topological) boundary of the set S, p. 27

© Springer International Publishing Switzerland 2016
L. Beck, *Elliptic Regularity Theory*, Lecture Notes of the Unione Matematica Italiana 19, DOI 10.1007/978-3-319-27485-0

$\mathbb{S}^{n-1}$	unit sphere ∂B_1 in $\mathbb{R}^n$, p. 26
x	generic point in $\mathbb{R}^n$, p. 2

Function spaces

$C(\Omega, \mathbb{R}^N)$	space of continuous functions on Ω, p. 2
$C_0(\Omega, \mathbb{R}^N)$	space of continuous functions on Ω with compact support, p. 2
$C^k(\Omega, \mathbb{R}^N)$	space of functions which are k-times continuously differentiable on Ω, p. 2
$C_0^k(\Omega, \mathbb{R}^N)$	space of functions in $C^k(\Omega, \mathbb{R}^N)$ with compact support in Ω, p. 2
$C^{k,\alpha}(\Omega, \mathbb{R}^N)$	space of functions in $C^k(\Omega, \mathbb{R}^N)$ with α-Hölder continuous k-th order derivatives, p. 3
$C^\infty(\Omega, \mathbb{R}^N)$	space of smooth functions on Ω, p. 2
$C_0^\infty(\Omega, \mathbb{R}^N)$	space of functions in $C^\infty(\Omega, \mathbb{R}^N)$ with compact support in Ω, p. 2
$L^p(\Omega, \mathbb{R}^N)$	Lebesgue space on Ω, integrable to the power p, p. 4
$\mathcal{L}^{p,\lambda}(\Omega, \mathbb{R}^N)$	Campanato space on Ω, p. 11
$L^{p,\lambda}(\Omega, \mathbb{R}^N)$	Morrey space on Ω, p. 11
$W^{k,p}(\Omega, \mathbb{R}^N)$	Sobolev space of integer order k on Ω, p. 19
$W_0^{k,p}(\Omega, \mathbb{R}^N)$	norm closure of $C_0^\infty(\Omega, \mathbb{R}^N)$ in $W^{k,p}(\Omega, \mathbb{R}^N)$, p. 19
$W^{k+\theta,p}(\Omega, \mathbb{R}^N)$	Sobolev space of fractional order $k+\theta$ on Ω, p. 22

Functions and operations on functions

$\mathbb{1}_S$	characteristic function of the set S, p. 24
$\operatorname{div} F$	divergence of a vector field F, p. 57
$E(f; x_0, r)$	excess of a function f in the ball $B_r(x_0)$, p. 100
$\exp$	exponential function, p. 13
Ef	extension of f to a larger set, p. 21
$(f)_S$	mean value of f on the set S, p. 6
f_-	negative part $-\min\{f, 0\}$ of f, p. 61
f_+	positive part $\max\{f, 0\}$ of f, p. 61
f^κ	κ-th component function of a vector-valued function f, p. 26
$D_i f$	(weak) partial derivative of f with respect to x_i, p. 2
Df	(weak) derivative of f, p. 2
$D^\beta f$	partial derivative $D_1^{\beta_1} \ldots D_n^{\beta_n} f$ for a multiindex $\beta \in \mathbb{N}_0^n$, p. 2
$D^k f$	the set of all partial derivatives of f of order k, p. 2
$\log$	logarithmic function, p. 25
$M(x_0, r)$	supremum of the relevant function in the ball $B_r(x_0)$, p. 66
$m(x_0, r)$	infimum of the relevant function in the ball $B_r(x_0)$, p. 66
$\operatorname{osc}(x_0, r)$	oscillation of the relevant function in the ball $B_r(x_0)$, p. 66
$\operatorname{Reg}_\alpha(f)$	set of points in which f is locally α-Hölder continuous, p. 99

$\mathrm{Sing}_\alpha(f)$	complement of $\mathrm{Reg}_\alpha(f)$, p. 99
spt f	support of f, p. 2
$\tau_{e,h}$	finite difference operator with respect to direction e with stepsize h, p. 26
$\triangle_{e,h}$	finite difference quotient operator with respect to direction e with stepsize h, p. 26

Functional analysis

X^*	dual space to a normed vector space X, p. 173
$\langle\,\cdot\,,\cdot\,\rangle_{X,X^*}$	duality pairing, p. 173
$x_j \to x$	a sequence $(x_j)_{j\in\mathbb{N}}$ in a normed vector space X converges strongly to x, p. 172
$x_j \rightharpoonup x$	a sequence $(x_j)_{j\in\mathbb{N}}$ in a normed vector space X converges weakly to x, p. 174

Other notation

$|\beta|$	length of a multiindex $\beta \in \mathbb{N}_0^n$, p. 2
δ_{ij}	Kronecker delta, defined as 1 for $i = j$ and as 0 for $i \neq j$, p. 26
N	functions usually take values in $\mathbb{R}$ (scalar case) or in $\mathbb{R}^N$ (vectorial case), p. 2
p'	conjugate exponent to $p \in [1,\infty]$, with $1/p + 1/p' = 1$ (convention $1/\infty = 0$), p. 7
p^*	Sobolev exponent to $p \in [1,n)$, i.e. $p^* = np/(n-p)$, p. 33

References

1. R.A. Adams, *Sobolev Spaces* (Academic, New York, 1975)
2. F.J. Almgren Jr., Existence and regularity almost everywhere of solutions to elliptic variational problems among surfaces of varying topological type and singularity structure. Ann. Math. **87**(2), 321–391 (1968)
3. H.W. Alt, *Lineare Funktionalanalysis* (Springer, Berlin, 2011)
4. L. Beck, Regularity versus singularity for weak solutions to elliptic systems in two dimensions. Adv. Calc. Var. **6**(4), 415–432 (2013)
5. L. Beck, T. Schmidt, On the Dirichlet problem for variational integrals in BV. J. Reine Angew. Math. **674**, 113–194 (2013)
6. H. Brezis, *Functional Analysis, Sobolev Spaces and Partial Differential Equations.* Universitext (Springer, New York, 2011)
7. R. Caccioppoli, Limitazioni integrali per le soluzioni di un'equazione lineare ellittica a derivate parziali. Giorn. Mat. Battagl. IV. Ser. **80**, 186–212 (1951)
8. S. Campanato, Proprietà di Hölderianità di alcune classi di funzioni. Ann. Sc. Norm. Super. Pisa Ser. III **17**, 175–188 (1963)
9. S. Campanato, *Equazioni ellittiche del II° ordine e spazi* $\mathcal{L}^{(2,\lambda)}$. Ann. Mat. Pura Appl., IV. Ser. **69**, 321–381 (1965)
10. S. Campanato, Differentiability of the solutions of nonlinear elliptic systems with natural growth. Ann. Mat. Pura Appl. Ser. 4 **131**, 75–106 (1982)
11. S. Campanato, Hölder continuity and partial Hölder continuity results for $W^{1,q}$-solutions of non-linear elliptic systems with controlled growth. Rend. Sem. Mat. Fis. Milano **52**, 435–472 (1982)
12. S. Campanato, Elliptic systems with non-linearity q greater or equal to two. Regularity of the solution of the Dirichlet problem. Ann. Mat. Pura Appl. Ser. 4 **147**, 117–150 (1987)
13. B. Dacorogna, *Direct Methods in the Calculus of Variations*, 2nd edn. Applied Mathematical Sciences, vol. 78 (Springer, New York, 2008)
14. E. De Giorgi, *Frontiere orientate di misura minima.* Seminario di Matematica della Scuola Normale Superiore di Pisa (Editrice Tecnico Scientifica, Pisa, 1960–1961)
15. E. De Giorgi, Sulla differenziabilità e l'analiticità delle estremali degli integrali multipli regolari. Mem. Accad. Sci. Torino Ser. III **3**, 25–43 (1957)
16. E. De Giorgi, Un esempio di estremali discontinue per un problema variazionale di tipo ellittico. Boll. Unione Mat. Ital. IV. **1**, 135–137 (1968)

© Springer International Publishing Switzerland 2016
L. Beck, *Elliptic Regularity Theory*, Lecture Notes of the Unione Matematica Italiana 19, DOI 10.1007/978-3-319-27485-0

17. J. Deny, J.L. Lions, Les espaces du type de Beppo Levi. Ann. Inst. Fourier Grenob. **5**, 305–370 (1953)

18. R.A. DeVore, R.C. Sharpley, Maximal functions measuring smoothness. Mem. Am. Math. Soc. **47**(293), viii+115 (1984)

19. E. Di Nezza, G. Palatucci, E. Valdinoci, Hitchhiker's guide to the fractional Sobolev spaces. Bull. Sci. Math. **136**(5), 521–573 (2012)

20. F. Duzaar, *Partielle Differentialgleichungen*. Lecture Notes (University of Erlangen, 2003)

21. F. Duzaar, J.F. Grotowski, Optimal interior partial regularity for nonlinear elliptic systems: the method of A-harmonic approximation. Manuscr. Math. **103**, 267–298 (2000)

22. F. Duzaar, G. Mingione, Harmonic type approximation lemmas. J. Math. Anal. Appl. **352**(1), 301–335 (2009)

23. F. Duzaar, Optimal interior and boundary regularity for almost minimizers to elliptic variational integrals. J. Reine Angew. Math. **546**, 73–138 (2002)

24. L.C. Evans, *Partial Differential Equations* (American Mathematical Society, Providence, 1998)

25. L.C. Evans, R.F. Gariepy, *Measure Theory and Fine Properties of Functions*. Studies in Advanced Mathematics (CRC, Boca Raton, 1992), pp. viii

26. J. Frehse, A note on the Hölder continuity of solutions of variational problems. Abh. Math. Sem. Univ. Hambg. **43**, 59–63 (1975)

27. N. Fusco, J.E. Hutchinson, $C^{1,\alpha}$ partial regularity of functions minimising quasiconvex integrals. Manuscr. Math. **54**, 121–143 (1985)

28. E. Gagliardo, Caratterizzazioni delle tracce sulla frontiera relative ad alcune classi di funzioni in n variabili. Rend. Sem. Mat. Univ. Padova **27**, 284–305 (1957)

29. F.W. Gehring, The L^p-integrability of the partial derivatives of a quasiconformal mapping. Acta Math. **130**, 265–277 (1973)

30. M. Giaquinta, *Multiple Integrals in the Calculus of Variations and Nonlinear Elliptic Systems* (Princeton University Press, Princeton, 1983)

31. M. Giaquinta, E. Giusti, Nonlinear elliptic systems with quadratic growth. Manuscr. Math. **24**, 323–349 (1978)

32. M. Giaquinta, E. Giusti, On the regularity of the minima of variational integrals. Acta Math. **148**, 31–46 (1982)

33. M. Giaquinta, E. Giusti, Differentiability of minima of non-differentiable functionals. Invent. Math. **72**, 285–298 (1983)

34. M. Giaquinta, E. Giusti, Quasi-minima. Ann. Inst. H. Poincaré Anal. Non Linéaire **1**(2), 79–107 (1984)

35. M. Giaquinta, L. Martinazzi, *An Introduction to the Regularity Theory for Elliptic Systems, Harmonic Maps and Minimal Graphs*, 2nd edn. Lecture Notes. Scuola Normale Superiore di Pisa (New Series), vol. 11 (Edizioni della Normale, Pisa, 2012)

36. M. Giaquinta, G. Modica, Almost-everywhere regularity results for solutions of non linear elliptic systems. Manuscr. Math. **28**, 109–158 (1979)

37. M. Giaquinta, G. Modica, Regularity results for some classes of higher order nonlinear elliptic systems. J. Reine Angew. Math. **311/312**, 145–169 (1979)

38. M. Giaquinta, G. Modica, Partial regularity of minimizers of quasiconvex integrals. Ann. Inst. H. Poincaré Anal. Non Linéaire **3**, 185–208 (1986)

39. D. Gilbarg, N.S. Trudinger, *Elliptic Partial Differential Equations of Second Order* (Springer, Berlin/Heidelberg/New York, 1977)

40. E. Giusti, *Direct Methods in the Calculus of Variation* (World Scientific, Singapore, 2003)

41. E. Giusti, M. Miranda, Sulla regolarità delle soluzioni deboli di una classe di sistemi ellitici quasi-lineari. Arch. Ration. Mech. Anal. **31**, 173–184 (1968)

42. E. Giusti, M. Miranda, Un esempio di soluzioni discontinue per un problema di minimo relativo ad un integrale regolare del calcolo delle variazioni. Boll. Unione Mat. Ital. IV. Ser. **1**, 219–226 (1968)
43. J.F. Grotowski, Boundary regularity results for nonlinear elliptic systems. Calc. Var. Partial Differ. Eqns. **15**, 353–388 (2002)
44. P. Hajłasz, Sobolev spaces on an arbitrary metric space. Potential Anal. **5**(4), 403–415 (1996)
45. C. Hamburger, A new partial regularity proof for solutions of nonlinear elliptic systems. Manuscr. Math. **95**(1), 11–31 (1998)
46. W. Hao, S. Leonardi, J. Nečas, An example of irregular solution to a nonlinear Euler-Lagrange elliptic system with real analytic coefficients. Ann. Sc. Norm. Super. Pisa Cl. Sci. IV. Ser. **23**, 57–67 (1996)
47. D. Hilbert, Mathematische Probleme. Vortrag, gehalten auf dem internationalen Mathematiker-Kongreß zu Paris 1900. Arch. der Math. u. Phys. **3**, 44–63, 213–237 (1901)
48. P.-A. Ivert, Regularitätsuntersuchungen von Lösungen elliptischer Systeme von quasilinearen Differentialgleichungen zweiter Ordnung. Manuscr. Math. **30**, 53–88 (1979)
49. T. Iwaniec, Weak minima of variational integrals. J. Reine Angew. Math. **454**, 143–161 (1994)
50. F. John, L. Nirenberg, On functions of bounded mean oscillation. Commun. Pure Appl. Math. **14**, 415–426 (1961)
51. J. Kristensen, G. Mingione, The singular set of ω-minima. Arch. Ration. Mech. Anal. **177**, 93–114 (2005)
52. J. Kristensen, G. Mingione, The singular set of minima of integral functionals. Arch. Ration. Mech. Anal. **180**(3), 331–398 (2006)
53. O.A. Ladyzhenskaya, N.N. Ural'tseva, *Linear and Quasilinear Elliptic Equations*. Mathematics in Science and Engineering, vol. 46 (Academic, New York/London, 1968), XVIII, 495p
54. G. Leoni, *A First Course in Sobolev Spaces*. Graduate Studies in Mathematics, vol. 105 (American Mathematical Society, Providence, 2009)
55. J.L. Lewis, On very weak solutions of certain elliptic systems. Commun. Partial Differ. Eqn. **18**(9–10), 1515–1537 (1993)
56. J.L. Lions, *Quelques méthodes de résolution des problèmes aux limites non linéaires* (Dunod, Gauthier-Villars, Paris, 1969)
57. J. Malý, W. Ziemer, *Fine Regularity of Solutions of Elliptic Partial Differential Equations*. Mathematical Surveys and Monographs, vol. 51 (American Mathematical Society, Providence, 1997)
58. V.G. Maz'ya, Examples of nonregular solutions of quasilinear elliptic equations with analytic coefficients. Funct. Anal. Appl. **2**, 230–234 (1968)
59. V.G. Maz'ya, S.V. Poborchi, *Differentiable Functions on Bad Domains* (World Scientific, River Edge, 1997)
60. N.G. Meyers, Mean oscillation over cubes and Hölder continuity. Proc. Am. Math. Soc. **15**, 717–721 (1964)
61. N.G. Meyers, J. Serrin, $H = W$. Proc. Nat. Acad. Sci. U.S.A. **51**, 1055–1056 (1964)
62. G. Mingione, Bounds for the singular set of solutions to non linear elliptic systems. Calc. Var. Partial Differ. Eqns. **18**(4), 373–400 (2003)
63. G. Mingione, The singular set of solutions to non-differentiable elliptic systems. Arch. Ration. Mech. Anal. **166**, 287–301 (2003)
64. G. Mingione, Regularity of minima: an invitation to the dark side of the calculus of variations. Appl. Math. **51**(4), 355–425 (2006)
65. G. Mingione, Singularities of minima: a walk on the wild side of the calculus of variations. J. Glob. Optim. **40**(1–3), 209–223 (2008)

66. C.B. Morrey, Partial regularity results for non-linear elliptic systems. J. Math. Mech. **17**, 649–670 (1967/1968)

67. J. Moser, A new proof of De Giorgi's theorem concerning the regularity problem for elliptic differential equations. Commun. Pure Appl. Math. **13**, 457–468 (1960)

68. J. Moser, A sharp form of an inequality by N. Trudinger. Indiana Univ. Math. J. **20**, 1077–1092 (1971)

69. J. Nash, Continuity of solutions of parabolic and elliptic equations. Am. J. Math. **80**, 931–954 (1958)

70. J. Necas, Example of an irregular solution to a nonlinear elliptic system with analytic coefficients and conditions for regularity, in *Theory of Nonlinear Operators Proceedings of the Fourth International Summer School, 1975* (Academic Sciences, Berlin, 1977), pp. 197–206

71. J. Necas, Counterexample to the regularity of weak solution of elliptic systems. Commentat. Math. Univ. Carol. **21**, 145–154 (1980)

72. L. Nirenberg, Remarks on strongly elliptic partial differential equations. Commun. Pure Appl. Math. **8**, 649–675 (1955)

73. D. Phillips, A minimization problem and the regularity of solutions in the presence of a free boundary. Indiana Univ. Math. J. **32**, 1–17 (1983)

74. T. Schmidt, A simple partial regularity proof for minimizers of variational integrals. NoDEA Nonlinear Differ. Eqns. Appl. **16**(1), 109–129 (2009)

75. J. Serrin, Local behavior of solutions of quasi-linear equations. Acta Math. **111**, 247–302 (1964)

76. L. Simon, *Lectures on Geometric Measure Theory* (Australian National University Press, Canberra, 1983)

77. L. Simon, *Theorems on Regularity and Singularity of Energy Minimizing Maps* (Birkhäuser-Verlag, Basel/Boston/Berlin, 1996)

78. G. Stampacchia, Le problème de Dirichlet pour les équations elliptiques du second ordre à coefficients discontinus. Ann. Inst. Fourier Grenob. **15**(1), 189–258 (1965)

79. E.M. Stein, *Singular Integrals and Differentiability Properties of Functions.* Princeton Mathematical Series, vol. 30 (Princeton University Press, Princeton, 1970)

80. M. Struwe, Funktionalanalysis I und II. Lecture Notes (2008). Available at http://www.math.ethz.ch/~struwe/

81. N.S. Trudinger, On Harnack type inequalities and their application to quasilinear elliptic equations. Commun. Pure Appl. Math. **20**, 721–747 (1967)

82. N.S. Trudinger, On imbeddings into Orlicz spaces and some applications. J. Math. Mech. **17**, 473–483 (1967)

83. K.-O. Widman, Hölder continuity of solutions of elliptic systems. Manuscr. Math. **5**, 299–308 (1971)

84. K. Yosida, *Functional Analysis.* Classics in Mathematics (Springer, Berlin, 1995)

85. W.P. Ziemer, *Weakly Differentiable Functions.* Graduate Texts in Mathematics, vol. 120 (Springer, New York, 1989)

Index

© Springer International Publishing Switzerland 2016

L. Beck, *Elliptic Regularity Theory*, Lecture Notes of the Unione
Matematica Italiana 19, DOI 10.1007/978-3-319-27485-0

LECTURE NOTES OF THE UNIONE MATEMATICA ITALIANA

Editor in Chief: Ciro Ciliberto and Susanna Terracini

Editorial Policy

1. The UMI Lecture Notes aim to report new developments in all areas of mathematics and their applications - quickly, informally and at a high level. Mathematical texts analysing new developments in modelling and numerical simulation are also welcome.

2. Manuscripts should be submitted to
Redazione Lecture Notes U.M.I.
umi@dm.unibo.it
and possibly to one of the editors of the Board informing, in this case, the Redazione about the submission. In general, manuscripts will be sent out to external referees for evaluation. If a decision cannot yet be reached on the basis of the first 2 reports, further referees may be contacted. The author will be informed of this. A final decision to publish can be made only on the basis of the complete manuscript, however a refereeing process leading to a preliminary decision can be based on a pre-final or incomplete manuscript. The strict minimum amount of material that will be considered should include a detailed outline describing the planned contents of each chapter, a bibliography and several sample chapters.

3. Manuscripts should in general be submitted in English. Final manuscripts should contain at least 100 pages of mathematical text and should always include

   – a table of contents;
   – an informative introduction, with adequate motivation and perhaps some historical remarks: it should be accessible to a
   reader not intimately familiar with the topic treated;
   – a subject index: as a rule this is genuinely helpful for the reader.

4. For evaluation purposes, please submit manuscripts in electronic form, preferably as pdf- or zipped ps-files. Authors are asked, if their manuscript is accepted for publication, to use the LaTeX2e style files available from Springer's web-server at
ftp://ftp.springer.de/pub/tex/latex/svmonot1/ for monographs
and at
ftp://ftp.springer.de/pub/tex/latex/svmultt1/ for multi-authored volumes

5. Authors receive a total of 50 free copies of their volume, but no royalties. They are entitled to a discount of 33.3% on the price of Springer books purchased for their personal use, if ordering directly from Springer.

6. Commitment to publish is made by letter of intent rather than by signing a formal contract. Springer-Verlag secures the copyright for each volume. Authors are free to reuse material contained in their LNM volumes in later publications: A brief written (or e-mail) request for formal permission is sufficient.

Printed in the United States
By Bookmasters